Agile Project Management

Methodology. A Comprehensive Beginner's Guide to Scrum, Kanban, XP, Crystal, FDD, DSDM

Sam Ryan

Table of Contents

Introduction

It's easy to dismiss a project manager's job and put it in a little stereotypical box painted in spreadsheet columns and rows that multiply to an endless infinity.

It's hard to actually understand the massive amount of work a project manager has to handle on a daily basis and the kind of pressure they have to face when they are torn between the team they spend eight hours every day with and a higher management level that is constantly asking for reports and profitability graphs.

If we have to be completely honest, project managers are modern-day heroes. They don't wear capes, and sometimes they can be downright annoying when they ask you about the status of your latest task. They don't have any other superpower aside from resilience and infinite patience. And they definitely don't have TV shows dedicated to them (although we would definitely back up a series based on the tumultuous life of an average PM in any software development company in the world).

Instead, project managers are frequently avoided and mistreated, misunderstood, and downright minimized. From the backbone of every company under the Sun (including *air flight companies,* mind you!) to spreadsheet carriers and email pushers, project managers seem to be in a constant limbo.

Truth be told, being a PM is difficult. Leaving aside the mad Excel skills you have to possess, being a PM is more about the people than it is about the reports—now more than ever, especially with agile project management taking over the world in a storm of variations.

The first installment of our Agile Project Management book was dedicated to exploring the fascinating world of agile and everything it has changed ever since its official inception in 2001.

Because we want you to have a solid foundation as you start reading the second installment of our Agile Project Management series, we will quickly move through the basic concepts we have already elaborated upon in the first book. If the volume at hand is where you stumbled upon us, we DO advise you to read the first one as well, as it will provide you with a treasure trove of information on where to start in the world of agile project management and how to tackle some of this framework's main challenges.

We started our first book with a quick history of agile—not because we like boring people in any way, but because we believe the history of agile project management is very tightly connected to the *needs* it came to solve. More specifically, agile project management was born on the back of a world that was speeding up like never before—a world that was completely incongruent with the strict limits of the traditional project management approaches practiced until then.

By the beginning of the 1990s, the first agile methods were slowly shaping up. Some of them had been used for decades already; others were born in an industry that was as new as the second half of the 20th century. In essence, they all came together under the Agile Manifesto at the beginning of 2001, and ever since then, they have continued to evolve and mix with each other (as you will see in the book at hand as well).

We also tackled the advantages of agile project management and why so many businesses are embracing it, either in its entirety or as part of a hybrid approach. From being able to deliver better quality to improving the quality of your team's life (at work and outside of it), agile project management has changed quite a lot. We are daring enough to say that agile has shaped up the world we currently live into a point most of us don't even know.

Just think of it: every single piece of technology developed from the 1990s onwards is largely based on one form of agile project management or another. Without agile, these products would have taken *a lot* more time to get to the market, and as such, our very lives would have been a little sadder without the existence of the internet, Wikipedia, and the infinite scrolls on Instagram every morning as we sip our cups of coffee.

Aside from the history and the advantages of agile, our first installment of this series also tackled a very important topic: how the basic Agile Principles (all twelve of them!) apply to real life and how they are translated into agile practices common across the entire spectrum of light-weight methodologies.

We know just how difficult the move from traditional waterfall approaches to agile project management can be. And, as such, we dedicated an entire chapter of our first book to showing you how to efficiently implement this framework into your business, regardless of what it may focus on.

We also tackled the topic of project management tools and provided readers with a brief analysis of some of the most popular tools one can use when embracing agile project management.

Finally, we dove deeper into two of the most complex subjects related to project management: risk management and scaling. The first topic is difficult to grasp in the context of agile project management because many people think agile involves no planning at all (which is highly untrue, as you will see in this book as well). The second topic is difficult because there are specific methods and best practices you should make sure to employ when aiming to scale agile, particularly in larger companies, where structures and documentation are already running through the very tissue of their organizational practices.

Enough about our previous book, though.

What does this book bring forward?

We mostly want to focus on the specific agile project management methodologies used by companies these days. While encompassing *everything* in the book is definitely not feasible, we aim to provide you with enough information to help you decide which of these approaches is best for your business.

We don't want to be repetitive, so we made sure to select the most precious bits of information for you. However, we do not claim the volume at hand to be a one-size-fits-all "recipe book." Au contraire: if you want this book to give you specific instructions on which method is the better choice, then you should probably put it aside.

We want to be completely honest with you (as we were in the first installment of this book). We cannot promise you the world and not deliver, and we genuinely believe that any kind of resource that will promise to give you the *ultimate* solution to anything is a complete and utter lie.

There's no such thing, regardless of whether you are talking about project management, apple pies, or efficient workouts.

Everyone is different. Every company is different. So, instead of chasing *ultimate solutions*, we encourage you to chase something more challenging but equally rewarding: *your* solution.

We want to empower you to make your own decisions when it comes to which agile methodology is best for you. As such, this book aims to give you the basic information on six of some of the most popular approaches out there: the definitions, the inner works, the principles, best practices, and all the other features that define them.

Our goal is for you to have the full picture of the agile project management world by the end of this volume. We want you to be able to analyze all the information we have provided both in this book and in our previous installment of the same series, and we want you to make the decision that works best for you, for your team, and, ultimately, for your organization.

As such, the first chapter here will be dedicated entirely to Scrum project management. As one of the two most popular methodologies, Scrum is frequently touted in mainstream business media as a receptacle of everything good in agile. While it is definitely full of advantages, Scrum should be perceived at its true value.

As such, we aim to provide you with a comprehensive overview of what Scrum is, how it relates to the general agile project management approach, how it connects to other agile methods, and how it functions. We will tackle Scrum teams, Scrum Ceremonies, Scrum Artifacts, and Scrum rules, so that you can gain the bird's eye view on how Scrum project management is actually practiced.

Our second chapter will be all about Kanban and lean. As a method that originated long before all the other popular agile approaches these days, Kanban is an interesting approach in the world of agile—one that did not pertain to the software development industry to begin with but which has become an integral part of it over the past two decades.

We will analyze the main Principles behind Kanban and lean project management, as well as the best practices that make these two connected approaches so unique and so useful for businesses from such a wide range of industries.

Further on, we will also analyze Kanban in relation to Scrum, provide you with a list of benefits Kanban brings along, and show you how Kanban can be implemented with the help of a popular project management tool.

Next, we will move on to another popular project management approach, one that is more exclusive to the software development world than many others are these days: Extreme Programming (XP).

We will take a look at what XP entails, from the Planning Game to the Coding Standards and the importance of the Metaphor in the entire Extreme Programming paradigm. By the end of this chapter, we hope you will have a full view of what XP is and how to incorporate it in your agile approach (in its pure form or as a hybrid between other agile and nonagile frameworks).

Our fourth chapter starts a series of analyses on one of the less popular, but equally valuable agile methodologies out there: The Crystal Method. You will be surprised to learn that despite the lack of popularity Crystal shows these days, it was one of the first ones developed in the 1990s (and it was also created within one of the most preeminent IT companies in the world!).

Beyond the debatable (lack of) popularity Crystal shows today, in 2019, this method has plenty to offer, its absolute flexibility being one of the most attractive features, even for larger organizations (like the one it originated in, actually).

Our fifth chapter in the book at hand is all about Feature-Driven Development (FDD), an agile method that might not sound very popular but which has been consistently incorporated with other agile approaches (including Scrum and Kanban). We will take a look at what FDD is and the elements that define it as a stand-alone methodology (elements that can be borrowed by other methodologies as well).

Last, but not least, we will approach one of the most apparently fearsome and misunderstood agile methods of the moment: The Dynamic System Development Method (DSDM). Frequently left aside because it feels too intricate and strict, DSDM can still provide you with plenty of value, particularly if you work in a company that is focused on structures and documentation and where a more relaxed agile approach would not function very well.

It would be remiss of us to think this book encompasses everything all these agile methodologies are. There is a very good reason there are so many books and so many materials and courses on these topics: they are wide and complex, much wider and much more complex than what many of the articles on Google would want you to think.

Truth be told, none of the methodologies we will approach in this book is actually *easy*, Agile itself is not easy, despite its apparent lightheartedness. Agile takes obsessive discipline, it takes a good strategist behind it, and it takes a lot of determination to function.

More than anything, agile takes a complete change in mindset. In agile projects, people do their jobs because it's the right thing, because this is what they are paid for, and because they want to be part of a successful story at the end of the project. They don't do it all because there's a "monster" called project manager that whispers "deadline" in their ear all the time.

In agile, things are done because they need to be done for the sake of evolution. People working in agile frameworks are more likely to have personal success stories in their résumé precisely because they get used to a mentality of hard work, dedication, and complete honesty. They know the path to growth is never easy and that it takes multiple increments to actually make it through, and they know this rule applies not only to whatever their job is but to their personal lives as well.

Agile has changed the way we think about success. It may not be fully apparent now, but if you look at all the inspirational business stories, at all the athletes and all the artists that won the game, at all the lovely stories of beauty and success the world has offered us, *agile* was always there.

They may not have called it that. In fact, it is quite likely that they didn't even think of a project management framework when they proceeded on their journey. But all these people you see on TV know that success is based on:

- honesty with oneself
- dedication
- hard work
- iterative and incremental implementation of growth
- constant feedback

At the end of the day, this is what agile is all about, regardless of whether you look at the oddities of Scrum and its Ceremonies or the apparent strictness of DSDM. And that is the main idea behind this book as well: there is no right or wrong agile approach. But some may just not be a good fit for you, and that's all right.

We hope the information lying ahead of you will not give you a headache. Instead, we hope it will open your mind (and, ultimately, your heart) to a world of opportunities in terms of what agile is (and what agile isn't).

Good luck and happy learning ahead!

Chapter 1: Agile Scrum Methodology

Agile project management is a complex topic, and one we have discussed quite in-depth over the course of our first book.

In (very) short, agile project management is all about keeping things flexible: from the way you plan ahead to the way you adapt to all the changes along the way. In turn, this helps deliver better products, in a shorter amount of time, and within a more acceptable budget.

As was shown in the first installment of this book series, when the official foundation of agile project management was laid back at the beginning of the 2000s, there was a real need for it. Software companies were experiencing massive lag in their deliveries of the final product precisely because the management system was not properly adapted to an industry where change is constant.

Agile project management has evolved a lot, and it has managed to incorporate a variety of directions meant to suit different types of companies, industries, and specific situations. Very often, organizations choose to work on hybrid frameworks (including hybrids that combine traditional/waterfall methods with agile methods).

Of all the agile frameworks out there, Scrum is, by far, one of the single most popular ones. This chapter will tackle the basics of the Scrum methodology, from what it is to how it connects into the 12

Principles of Agile, how the team is structured, what artifacts are, and how to actually practice it.

What is Scrum Project Management?

When they hear "Scrum," most people think of the (in)famous Daily Standup meeting. From the outside, it looks like a bunch of people sitting in a circle and taking turns in standing up and talking for two minutes.

It is easy to understand why this might seem silly, forced, or downright annoying.

The absolute truth is that Scrum project management is about a lot more than just the Daily Standup.

Indeed, this is a central part of Scrum project management, one that is not silly at all, actually.

Beyond that, Scrum project management has a lot of components and each of them brings its own contribution to the benefits of agile Scrum.

There are many ways to define Scrum project management (including through the perspective of the popularity with which it has been used). One of the best ways to do it is by calling it an "agile framework" that assigns specific roles to each team member, focuses on regular meetings at different levels, and focuses on a very well-defined mindset.

Scrum has gained a lot of fame in a relatively short amount of time. From the early 90s when it was first developed to today, Scrum has managed to become not only one of the most popular agile methods but very often *the main* method people associate with the entire concept of "agile project management."

The popularity of Scrum is largely due to its ease of use. Everything about Scrum is meant to put people in a certain productivity mindset, rather than impose on them specific tactics. Although from the outside it might seem the other way around, the techniques specific to Scrum are not meant to seclude the mindset of those who participate in a project, quite the other way around actually

Studies show that 94% of companies used agile Scrum project management as part of their approach in 2017. Only 16% of them used Scrum exclusively, while the remaining 78% used Scrum in combination with other methods.

Scrum project management is, without a doubt, one of the best ways to ease your way into agile project management, precisely because Scrum is very close to the ideal agile projection of how a product should be planned, developed, tested, and delivered.

Beyond any other definition, though, what Scrum is truly defined by are its benefits. As mentioned before, all agile methods were born out of a dire need: that of delivering faster, better results.

In (very) short, these are some of the benefits of Scrum project management:

- Generating revenue. Because Scrum involves an incremental release of the product, it is far more likely that money will come back into the business sooner, rather than later.
- Providing better quality. The incremental development of the product ensures that it will be perfected with each iteration. As such, the end product will be of higher quality as well.
- Offering better transparency for everyone. The entire process is transparent for every stakeholder and team member, from beginning to end.
- Better risk management. The incremental nature of Scrum also helps Scrum Masters/project managers manage the associated risks with a lot more efficiency, as they will adapt the risk management plan along the way, according to the challenges they encounter.
- Better flexibility. Since this is agile (which means it is a project management approach that is both iterative and incremental), you will have better flexibility to adapt to changes.
- Better control of the costs. Again, this is due to the incremental nature of Scrum and how it allows you to manage the entire project more accurately.

- Better customer satisfaction. This ties back into the fact that Scrum allows you to release *chunks* of product and attract feedback from your customer. As a result, your final product will be better as well.
- Faster product release. When you can adapt the product along the way, you can actually release it sooner too.
- Better work environment. Scrum promotes a culture of cooperation, collaboration, communication, and a good work/life balance. All of these things lead to a better work environment, where people are allowed and encouraged to grow.

What is Scrum in Relation to Agile Project Management?

Scrum project management is so tightly connected to the world of agile project management that the two have become almost synonymous in the collective mind.

Agile and Scrum are so tightly connected that when someone says "agile," they almost instantaneously think of practices that are very specific to Scrum, like the Daily Standup we mentioned before, the Scrum Master, or the Scrum board, for example.

Indeed, if you look at the benefits of general agile project management and Scrum project management, you will see how close they are in nature. In most cases, the latter is just a practical, "regulated" mirror of the philosophy the first tries to bring forward.

Even with all the closeness between the two concepts, it is still very important to make sure you understand that they are quite different.

If you want to picture it, imagine a tree that makes apples of different colors. The tree is the agile project management approach, while the colored apples are the different methodologies (which we will approach throughout this book). Scrum is one of those apples. While it grows from the same tree and follows the same general rules on how apples grow, it has chosen to have its own unique color.

Agile project management is an approach that focuses on iterative development, while Scrum is a branch that has created its own rules of how to play the game by the general Agile Principles as they were laid down in 2001.

In the end, you cannot actually compare agile project management and Scrum. One is an entire category of approaches, while the other one is a specific approach. Indeed, it is an approach that gets very, very close to the ideological and methodological basis of agile project management.

However, when you discuss agile project management and Scrum in relation to each other, you should do it from the perspective of evolution, growth, and branching, rather than from the perspective of comparison per se.

What is the Scrum Methodology Compared to Other Agile Approaches?

The world of agile is quite large and fascinating, and, unlike the world of waterfall, it is a world full of nuances.

The vast majority of companies don't take *one* methodology to heart. Rather than that, they create their own mixes, according to their specific needs and their specific company policies.

Scrum is, as it was mentioned before, one of the most popular agile methodologies, so much so that it is almost synonymous with "agile" in general. However, that doesn't make the other methodologies less agile; it just makes them different.

At their foundation, all agile methodologies spring from the 12 Agile Principles, but each of them shows its own particularities.

For instance, when you discuss Scrum and Kanban, you might feel that you are discussing the same methodology. In reality, however, they are quite different in how they both approach and tackle the Agile Manifesto's dictums.

The differences between Scrum and other agile methodologies will become clearer as you move through this book. However, let us briefly dive into a short comparison that will give you a better idea on how Scrum relates to the other popular agile methodologies.

Scrum vs. Kanban

As mentioned above, the two are quite similar in the sense that:

- Both abide to the general agile Principles.
- In both cases, the work in progress is limited (but Kanban emphasizes more on this).
- In both cases, the work is broken down in small increments.

- In both cases, work is scheduled using the "pull" method, rather than the "push" one.
- In both cases, the teams are organized.
- Both methods emphasize transparency.

As for the differences between the two, they include the following:

- Scrum uses time-boxed iterations, while Kanban doesn't.
- Scrum is all about fast-changed processes, while Kanban isn't.
- Scrum uses burn-down charts for each iteration, while Kanban doesn't.
- Scrum limits the work in progress through its sprint plan, while Kanban limits it through its workflow (each team member is allowed to handle only one task at a time).
- Kanban is generally less structured.
- Scrum assigns prescribed roles (Product Owner, Scrum Master, etc.).
- Scrum is all about getting work done faster, while Kanban is all about improving the process.

Scrum	Kanban
Timeboxed iterations prescribed	Timeboxed iterations optional. Can have separate cadences for planning, release, and process improvement. Can be event-driven instead of timeboxed.
Uses Velocity as default metric for planning and process improvement.	Uses Lead time as default metric for planning and process improvement.
Cross-functional teams prescribed.	Cross-functional teams optional. Specialist teams allowed.
Items must be broken down so they can be completed within 1 sprint.	No particular item size is prescribed.
Burndown chart prescribed	No particular type of diagram is prescribed.
WIP limited indirectly (per sprint)	WIP limited directly (per workflow state)
Estimation prescribed	Estimation optional
Cannot add items to ongoing iteration.	Can add new items whenever capacity is available
A sprint backlog is owned by one specific team	A kanban board may be shared by multiple teams or individuals
Prescribes 3 roles (PO/SM/Team)	Doesn't prescribe any roles
A Scrum board is reset between each sprint	A kanban board is persistent
Prescribes a prioritized product backlog	Prioritization is optional.

Scrum vs. Extreme Programming (XP)

Although XP is probably less popular than Scrum (and Kanban, for that matter), it is worth looking into how it relates to it.

Some of the main similarities between the two methodologies include the following:

- They are both agile.
- They both focus on delivering quality products in a short time span.
- Both start with a planning meeting.
- Both use sprints to organize the work.

As for the differences, they include the following:

- Scrum includes Daily Meetings, while XP doesn't.
- The length of each sprint is usually shorter in XP.
- The focus lies on delivering working software, rather than doing it by a specific product release date.
- Scrum doesn't allow changes to be made during a sprint, while XP does.
- XP focuses more on engineering principles (such as automated testing, test-driven development, pair programming, and so on).
- In Scrum, the meetings are coordinated by the Scrum Master, while in XP the team members take turns in coordinating meetings.

Scrum	XP
• Changes in sprint are not allowed • Once tasks for a certain sprint are set, the team determines the sequence in which they will develop the backlog items • The Scrum Master is responsible for what is done in the sprint, including the code that is written • The validation of the software is completed at the end of each sprint, at Sprint Review	• As long as the team hasn't started working on a particular feature, a new feature, of equivalent size can be swapped into the interation in exchange for an un-started feature • Tasks are taken in a strict priority order • Developers can modify or refactor parts of code as the need arises • The software needs to be validated at all time, to the extent that tests are written prior to the actual software

Scrum vs. Crystal

Scrum and Crystal are also worth comparing. In short, the main similarities between the two include the following:

- Neither puts a lot of emphasis on documentation.
- Both abide by the general agile Principles.

There are some more than notable differences as well:

- In Scrum, the client is represented by the Product Owner, while in Crystal, the customer is involved in the process.
- Crystal does not connect planning and development to specific requirements like Scrum does.
- Crystal focuses even more on face-to-face meetings.
- Crystal focuses more heavily on the *team* members and adapts the processes according to them.

Scrum vs. Feature-Driven Development (FDD)

Feature-Driven Development (FDD) is still one of the most extremely valuable agile methodologies out there. Although different in many ways, it does show some similarities with Scrum, such as:

- They are both agile methodologies.
- Both of them follow the same basic steps.
- They can act as a complement to each other.

The differences between FDD and Scrum include the following:

- In FDD, there is no Scrum Master, but chief programmers who act as leaders and mentors.
- FDD is "ultra-light" as compared to Scrum, as it is far less prescriptive and more adaptive.
- Scrum does not recommend any specific engineering practice.
- FDD shows a longer feedback loop.
- Scrum focuses on self-organized teams more than FDD.

Scrum vs. Dynamic System Development Method (DSDM)

Same as with the other methodologies we have touched upon in this comparative section, Scrum and DSDM show both similarities and differences.

Some of the most important similarities between Scrum and DSDM include the following:

- Both of them abide by the general agile Principles.
- Both of them can work when they are combined.

As for the differences, the most notable ones include the following:

- DSDM tends to be even more prescriptive in terms of team roles.
- In the case of DSDM, all the basic information has to be set from the beginning: features, quality, time, and cost.
- DSDM tends to be easier to adapt to a corporate environment.

This is, of course, a very brief overview of how Scrum and other agile project management methodologies are similar and different at the same time. As we go more in depth on each method, the similarities and the differences will become more obvious.

The Scrum Team

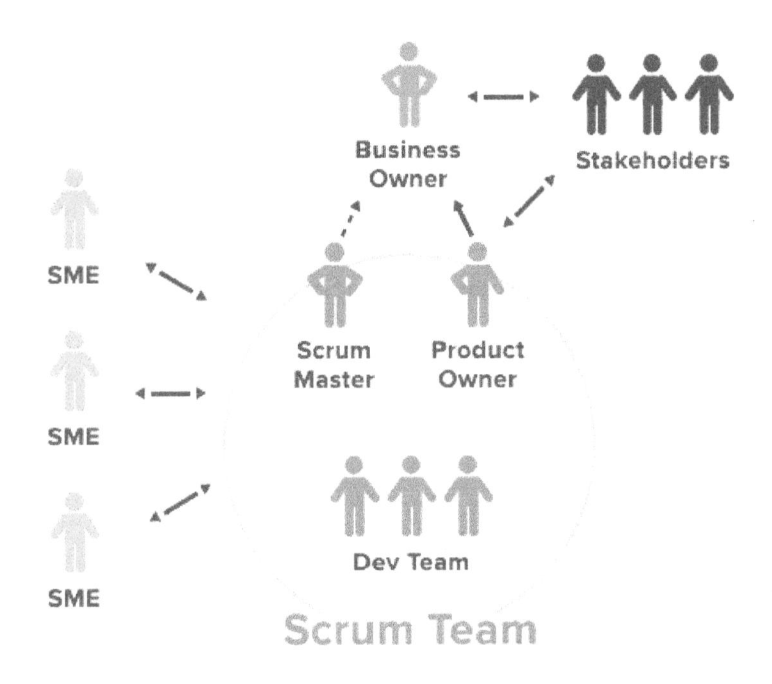

The Scrum team lies at the very core of the entire methodology. It is where Scrum starts and where the true magic happens too. It is also one of the Scrum features everyone from the outside can easily spot (let's face it, adding Scrum Master to your resume sounds way cooler than other titles, doesn't it?).

Beyond the buzz and the "glam" of what the Scrum team feels and looks like from the outside, it is more than worth noting that in a

Scrum team, everyone has their own part to play. Everyone's role is quite clearly defined, and everyone comes together to fight for the common goal: delivering the project on time, within the limits of the budget, and at the highest standard of quality possible as well.

In general, a Scrum team will be split as follows:

The Scrum Master

The Scrum Master's main role is that of planning and making sure the plan is followed through. Most of the times, this means he/she will have to:

- clear obstacles
- create an environment of efficiency
- address team dynamics
- create and maintain the communication and relationship between the Product Owner and the team
- protect the team from external interruptions

The Product Owner

At the confluence between a Scrum Master and a more traditional project manager, the Product Owner in a Scrum team has a pretty well-defined role as well. They are meant to:

- act as a liaison between the Team and the Customer

- act as a liaison between the Team and other Stakeholders
- help the Team estimate the size of the stories
- help the Team split larger stories into smaller chunks

The Team Members (Development Team)

This part is pretty straightforward: The Development Team is a group of people handling the actual development of the product. Some of these people might act as developers and others might act as Quality Assurance engineers, for example. But together, they form the "Team" in a Scrum environment, and together with their Scrum Master, they will coordinate themselves to create products that suit the user requirements and the information collected by the Product Owner from the customer.

Aside from the roles held in a Scrum team, it is also more than worth mentioning that there are a few guidelines all Scrum teams are meant to follow (and all of these guidelines are inspired by the agile Principles, of course).

Some of the most important rules in a Scrum team include the following:

- Everyone follows the same rules and works for a common goal.
- The Team (in its entirety) must be accountable for the product delivery.

- The Team must be empowered to work at its maximum efficiency.
- The Team must be autonomous and self-organized (as much as possible at least).
- The Team is usually quite small and there are no subteams under it.
- There must be a balance of skills within the team.

Depending on how "purely agile" and "purely Scrum" you want to be in your approach, you might also want to make sure that your team is located in the same place, so that meetings and collaboration can take place face to face every time.

Scrum Events (Ceremonies)

Aside from the predefined roles, Scrum also allocates very specific events (also called "Ceremonies") for very specific purposes.

There are four main Scrum Ceremonies, each with a very clear goal, as follows:

Sprint Planning

The Sprint Planning Ceremony is all about, well, *planning*. Since the entire project will be split into small iterations (usually lasting for a couple of weeks or a little more), every such iteration will start with a new planning session.

As such, a project will have as many Sprint Planning sessions as there are sprints. Furthermore, every Sprint Planning will last for about one or two hours, depending on how many issues there are to discuss.

It is of the utmost importance to include the entire team in every Sprint Planning, so that everyone can participate in the planning itself. This way, everyone will bring their own input to the table, and you will be able to create a more accurate plan for the Sprint ahead of you.

The Scrum Master, the Product Owner, and the Development Team all have to participate in the Sprint Planning if things are to be handled smoothly from thereon. DO encourage people to communicate if they are skeptical or doubtful about anything!

Daily Scrum

This is the most well-known Scrum Ceremony *ever*.

Almost everyone has heard about it.

But the truth is that most people completely misunderstand what the Daily Scrum is all about.

In short, the Daily Scrum is a short, 15-20-minute meeting held every day. The Scrum Master and the Development team are the main participants here. They gather in a circle and each of the members takes their turn standing up and talking for 2 minutes about:

- what they did the day before
- what they plan on doing today
- what bottlenecks they might have

Despite what many would believe, the Daily Scrum has nothing to do with micromanagement and constantly following your team to ensure they do their job. In fact, the Daily Scrum is meant to help team members be more productive by removing the bottlenecks from their path to achieving what they aim for every day.

Sprint Review

In this Ceremony, the Team demonstrates the work they have done throughout the duration of the sprint.

The Sprint Review does not have a time cap; it can last for as long as it is needed for the team to demonstrate the work, they have done throughout the sprint that ended.

Furthermore, keep in mind that it is quite important to maintain the Sprint Review as positive as possible. Even if mistakes were made and even if the sprint results are not what you aimed for, it is still essential to congratulate the team for what they did right.

This will help you maintain team morale, and it will help you step into the new sprint.

Sprint Retrospective

If every sprint starts with a Sprint Planning session, it must end with a Sprint Retrospective as well.

This is a time to bring the entire Team together and analyze what went right and what didn't go as right as it should have. It is a time to discuss, but it is important for these discussions to lead to actual action too.

Like all agile methodologies, Scrum is all about adapting to change, and the Sprint Retrospective is meant to help the Team incorporate the changes they need to make for the future sprint.

Scrum Artifacts

Aside from the predefined roles and the Ceremonies, Scrum also makes use of so-called "artifacts."

What these artifacts are, in fact, are tools used for better planning and development of a product. Although quite specific to Scrum itself, the most common artifacts in this agile project management methodology have been borrowed by other methodologies as well, precisely because they are so useful.

The most popular Scrum artifacts include the following:

Product Vision

This artifact describes the product in a nutshell. It should be something short and easy to remember—something that gives everyone a good direction and helps remind them of the higher goal behind their daily work.

Sprint Goal

This tool is used to describe the specific goal of each sprint. For instance, if you are developing a social media management platform, your first sprint's goal might be to create the basic platform upon which all the features will be built (i.e., a platform that connects to various social media channels).

Product and Sprint Backlog

These two artifacts are very frequently mistaken and considered to be one and the same, but they are quite different in essence.

In short, a product backlog is the entire collection of tasks that have to be handled throughout the development process (until the delivery of the final product). A sprint backlog will subtract those tasks that pertain to that specific sprint only, though.

Burn-Down Chart

This artifact is a chart the Scrum Master and the Team will use to observe the progress of the team's efficiency, as well as how realistic the initial plan was, over the course of the development process. The burn-down chart will be used to make adjustments on the go, as needed.

Increment

To define it simply, an increment is a Scrum artifact used to describe the totality of the Product Backlog items completed during a sprint (as well as all those that were completed in all the Sprints before it).

Aside from these artifacts, it is also very important to remember the fact that you absolutely *have to* set the definition of "done" with your team members. Believe it or not, people have different versions of what "done" actually is, so it is essential to clear things up with your team before you even start working.

Scrum Rules

The basic rules of Scrum are, in fact, the basic rules of agile. As long as you abide by the 12 Principles (as they were extensively described and explained in the first increment of our Agile Project Management book), and as long as you make use of the Scrum artifacts, Scrum Ceremonies, and Scrum roles, you can definitely call yourselves a Scrum team.

Aside from all that was mentioned so far, it is also worth noting some of the most important rules in Scrum project management:

- A sprint cannot last for more than four weeks (ideally, it will last for less than that).
- You shouldn't take breaks between the different sprints.
- You should make sure all sprints last for the same amount of time.
- The main underlying goal of every single sprint should be "workable" and potentially "usable" software that can be delivered to the customer as such.
- You shouldn't skip any of the Scrum Ceremonies (the Planning included).
- Most meetings should be time boxed, except for the Sprint Review.
- Your Daily Meeting should take place every day, at the same time of the day, and preferably in the same space.
- Sprint Reviews should include feedback from the stakeholders at all times, so that you know what conclusions to draw during the Sprint Retrospective.
- There should be no break between the Sprint Review and the Sprint Retrospective. The feedback received during the Review should be fresh for the Retrospective, so that the

entire team can discuss it and jot down the actionable steps to implement in future Sprints.

Practicing Scrum

In theory, Scrum sounds like actual fun because who doesn't like planning with their entire team, and who doesn't like not having to constantly draw attention to each and every member of the team to do their work?

In practice, Scrum might prove more difficult to implement, especially in a business that is already used to traditional project management methods. It all depends on how smoothly you want to introduce Scrum to your team and your higher management alike.

Some of the best practices you might want to incorporate in your approach include the following:

- Stakeholders should be an integral part of the entire development process. Invite them to some of your meetings, particularly Sprint Planning and preproject meetings.
- If you already have a team that is well formed, don't break it. It is far easier to introduce Scrum to them because they already know each other, they already speak each other's "language," and they already know how to properly communicate with each other.

- Team building is important. It doesn't matter if you choose Scrum, another agile methodology, or simply choose to remain traditional. Team building can really make a difference.
- Include the team in the Sprint Planning, especially when it comes to making estimates. We have already tackled the subject of how to play estimation "games" with your team in our first installment of this book series, so we definitely advise you to check that out. These games are not just fun; they also help you get actually better estimates of how long some tasks will take until completion.
- The Product Backlog and the Sprint Backlog should be delimited for better organization.
- Try to prioritize the items in the Product Backlog according to their importance and how dependent other items are on them.
- DO use a Scrum board. Typical columns in a Scrum board will include: Stories, Not Started, In Progress, Done, Blockers, In Testing, In Review (by the Product Owner, for example). These columns should be considered as separate from the Product Backlog and the Sprint Backlog.
- Calculate the velocity of your team with regularity. It will help you spot any efficiency problems while they are still young and easily repairable.

- Your developers and your testers should work together, and yes, this means they should be in the same room as well.
- All the bugs you find during a current sprint should be fixed in the following sprint.

Obviously, this is the very short version of what the best practices of Scrum look like. As mentioned at the beginning of this section, though, it is far more important to adhere to the general Principles than any prescriptivist methods.

Scrum can be amazingly useful, and it can definitely be a very good "gateway" into the world of agile in general. Because its practices and its particulars are much more popularly known among many people, your team and your higher management might be more open towards adopting Scrum (as opposed to other, more extreme, and far more lightweight methodologies).

Scrum tends to be a milder version of agile, and for this reason, it also tends to be combined with practices from other agile methodologies (as it was also mentioned when we compared Scrum and the other approaches we will tackle throughout this book).

We invite you to learn more about the other agile methodologies as well. Next chapter, we will discuss Kanban (the other, probably equally popular agile approach). Following that, we will dive a little deeper into the world of agile and take a closer look at XP

(Extreme Programming), Crystal, Feature-Driven Development (FDD), and the Dynamic System Development Method (DSDM), all of which are incrementally more agile, and all of which pose particulars that might be more suitable for your company.

In the end, you and your team will decide which of the methodologies works best for you or if you need to combine multiples of them to create the right “program” for your team’s best efficiency.

Chapter 2: Lean and Kanban Software Development

Kanban is, perhaps, the oldest project management method in the realm of agile. While it might have been officially adopted as part of the agile family when the Principles and the Manifesto were laid down, Kanban's history goes far beyond that.

The beginnings of Kanban date as far back as the 1940s, when Toyota designed the first Kanban system. Albeit far more rudimentary than what is nowadays known as the "Kanban agile project management methodology," the system created by Toyota back then survives to this date, and it has been borrowed by and adapted to a multitude of industries.

What started off as an approach that meant to minimize waste and improve efficiency is, these days, an entire methodology used in software programming, marketing, hospital management, and a variety of industries that are apparently as disconnected from each other as they can possibly get.

When you get beyond the differences between all these industries, though, you realize that all of them face the same kind of issues that pose the same kind of questions:

- How to reduce waste?
- How to maximize the efficiency of a team?
- How to make the work environment a healthier and more productive one?

A bit less prescriptive in nature than Scrum, Kanban manages to offer an answer to these questions (and more). As you will see throughout this chapter, Kanban and lean project management are very tightly connected, and they both fit under the agile project management umbrella quite well.

Without further ado, let's take a closer look at what Kanban is and how to use it in your company.

Main Principles of Lean Methodology

The lean methodology (or "lean thinking" as it is sometimes referred to) is not necessarily a project management approach, framework, or methodology in the fullest sense of the word.

Rather, this is a business philosophy that is adjacent to many of the agile project management methods out there.

The lean methodology relies on the Japanese concept of "Kaizen" (meaning "improvement"). So, right from the very beginning, you can see how tightly connected lean and agile are in essence. In its turn, Kaizen relies on five main principles:

Value

In Kaizen, providing value to the customer is the ultimate goal. However, in order to do that, you will first have to understand what "value" means for each of your customers. What makes them pay for your product, regardless of whether that is a car, a software program, or a piece of clothing? What triggers someone to buy the product and be happy with it?

Once you understand the kind of value you can offer your customer, you also understand how to minimize waste and how to create a product that is efficient both in terms of quality and in terms of price.

Value Stream

The Value Stream is the road map of the entire product, from idea to the end user. If you want to eliminate waste and provide your customer with genuine value, you have to understand every single step the product will take, from its raw materials to how the customer will use it.

Flow

In lean, your manufacturing/development flow is of the utmost importance because you have to constantly make sure there are no pauses in production. Every pause is a waste of time, and since eliminating waste is the beating heart of every lean approach, it is unpardonable not to ensure a smooth flow of the entire production process.

Pull

Eliminating waste also means that you will have to create a number of products that is *just enough*. Too much product will lead to waste, but a small amount of product will not satisfy the customer. As such, lean project management makes use of the *pull method* to ensure the production process is just in time and that the product (and number of products, in the case of the physical ones) is perfectly coordinated with the demand (of the market or of a specific customer).

Perfection

Same as in agile project management, Kaizen promotes the idea of *continuous improvement*. Its end goal is nothing less than perfection: a product that works perfectly, comes to the market at the perfect time, and fits the needs of the customer in a perfect way.

While absolute perfection might be hard to reach (if not impossible, in most cases), striving towards it will help you and your team *ask for more* from yourselves and continuously work towards an ideal goal.

Aside from the Kaizen principles that support lean project management, there are seven more principles that come as a complement to everything and circle back to the idea of providing true value to the customer. What is commonly known as the "Seven Lean Principles" includes the following:

- eliminate all waste (as much as possible)
- support and encourage learning and knowledge
- decide as late as possible
- deliver as soon as possible
- emphasize the power of the *humans* in your team
- believe in integrity and quality
- constantly optimize and observe the *whole*

These Seven Principles will be discussed throughout the next seven subsections of this chapter, so we will not dwell on them too much. It is important to know, however, that they connect very closely to both the Kaizen Principles and to the 12 Principles of Agile project management. Looking at the lean principles through these two points of view will allow you to understand where Kanban falls on the agile spectrum, and, ultimately, whether or not it is a good choice for your team and your organization.

Eliminating Waste

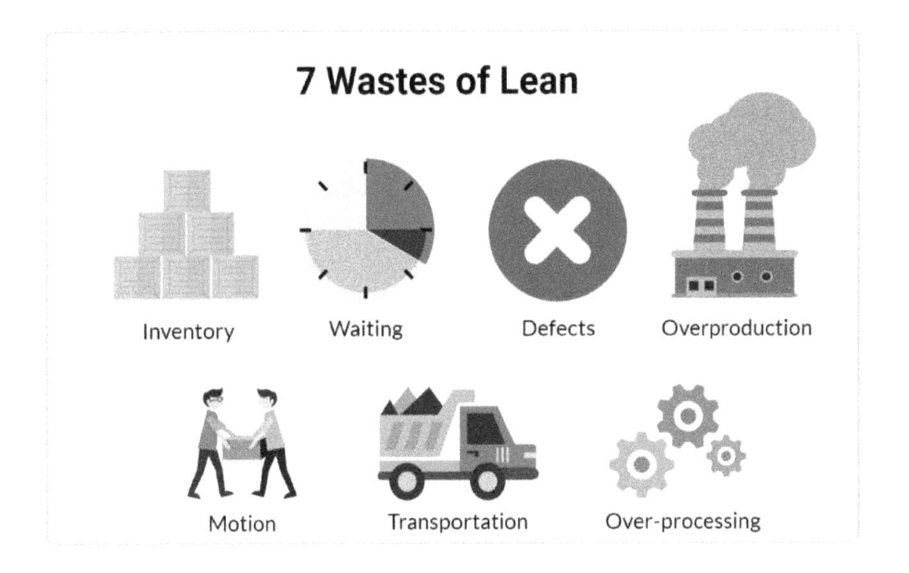

In lean project management, waste is the ultimate evil. You will never find an agile methodology to focus so much on this specific aspect of production, and this is mostly related to the fact that lean and Kanban both draw their roots from manufacturing environments, where waste is very palpable, physical, and, well, *costly*.

You might not be able to sense waste in software programming, not in the sense that you can hold it in your hand or that you can look at a pile of it as it collects in the corner of your office. You can, however, *feel* its effects, and this is where lean proves its efficiency in everything connected to software project management.

In the original form of lean and Kanban, the precursors of this approach identified eight main types of waste one should avoid at all times:

1. Transport
2. Inventory
3. Motion
4. Waiting
5. Overproduction
6. Overprocessing
7. Defects
8. Skills (also known as the 8th type of waste because it was not originally included in the core principles of lean)

These types of waste are quite self-explanatory when you look at them from the point of view of a manufacturing company. However, let's put them in the context of a software development company, so that you understand how they can still apply even when the final product is not a physical one:

1. Transport may refer to an unnecessary movement of information (e.g., too many reviewers involved in the

process, too many backs and forth emails passed between the different stakeholders, and so on).

2. Inventory may refer to accumulating too much information to be entered in the system at a later date or accumulating too much information that is not related to the other members of the team in due time.
3. Motion may refer to switching too many tools, moving too much between different workstations, and, in general, *moving* too much.
4. Waiting may refer to any kind of bottleneck that prevents one (or more) of your team members to move further with the development process.
5. Overproduction may refer to anything that ranges from hitting the "Reply All" button (when not everyone is interested) to simply delivering *much* faster than needed.
6. Overprocessing may refer to adding any kind of extra feature that was not requested initially.
7. Defects may refer to any type of bugs or issues the product might have.
8. Skills may refer to a lack of talent in your company (e.g., you are struggling to find senior developers).

As you can see, the same waste principles can be applied to manufacturing just as they can be applied to software programming. Likewise, you can transpose the same principles to any other industry: governmental institutions, educational institutions, health organizations, and so on.

Waste is a true demon, especially in a world that seems to be infatuated with wasting time, food, and resources on a daily basis. Fighting off this demon might not be easy, but as the decades of experience at Toyota prove, it is more than worth the effort.

Amplifying Learning

The basic concept behind this principle is quite simple: the more you and your team learn and build knowledge, the more efficient the process will be and the better the final product will be on all verticals (cost, delivery time, quality).

Just because the principle itself is simple to understand, it doesn't mean it is equally simple to apply in real life. In theory, acquiring knowledge always sounds easier than what it actually is in practice, and this is why you have to pay attention to *how* exactly this principle is brought into materialization.

Some of the main ways to amplify learning in an agile Kanban team include the following:

1. Constant feedback. Since this is one of the main tenets of agile project management in general, constant feedback should be seen as an absolute must. Both you and your team must get used to always giving and receiving feedback from each other and from your customer. Most

importantly though, you all need to make sure you actually *apply* that feedback as well (there is no point in giving and receiving feedback if you don't do anything about it).

2. Asking yourself "how" and "why." When you want to get down to the root of a problem, these two questions will lead the path. Imagine them as your guiding light through the harshest moments of your project. Do keep in mind that it is important to be honest about your answers. Also, do keep in mind that you should go in-depth with your answers as well; if needed, ask the same question multiple times to get to the root cause of the problem.
3. Pair Programming. Although this is not a practice inherent to Kanban, it is a useful exercise for your team members. Basically, what Pair Programming means is pairing your developers in groups of two. One will be the driver (the one who writes the code), and the other one will be the observer (the one who guides the driver).
4. Code reviews. Also known as "peer code reviews," this practice will help your team members give feedback to each other and learn from each other's experience.
5. Documentation and wiki. Although agile project management doesn't emphasize the importance of documentation, it doesn't mean there should be *none*. The focus should lie on working and delivering the product, of course. However, documentation and wikis (knowledge bases) can be passed on from one team member to another

(including new recruits who might not be completely familiar with the processes and practices you have employed so far).

6. Training sessions. Training, courses, and seminars will definitely help everyone grow. It is quite important to pick your subjects and your training carefully. You don't want to bore your team with information they have heard a thousand times before. Instead, you want these sessions to prove genuinely helpful for everyone.
7. Track your progress. Charts and project management tools can make a world of difference when it comes to tracking the progress of your team, both in terms of efficiency and in terms of quality.

These are some of the most common practices you can employ to accelerate the learning of your team. As mentioned in the beginning, the more they know, the more they will be able to deliver in terms of quality, efficiency, and speed.

Deciding as Late as Possible

This is one of the most poorly understood principles behind lean and Kanban. A lot of people believe this principle refers to postponing all decisions until it is too late (and thus, displaying an irresponsible behavior towards decision making).

In fact, this principle is about the exact opposite concept: keeping your options open. Making decisions too early in the process might affect the end result if you don't have all the data at hand. However, postponing the final decision until you have plenty of data to make an informed decision will help you get closer to what really needs to be done.

In other words, there is no point in planning for months in advance. As per the agile Principles, you should always be ready to adapt to change, and a rigorous, strict plan will not help with that.

This doesn't mean you shouldn't plan *at all*. However, you should do it in smaller increments, so that you can easily adapt along the way as per the information you receive throughout the development process.

Delivering as Fast as Possible

Again, this is a frequently misunderstood concept, not just in lean and Kanban, but in agile project management in general. Delivering ahead of the schedule is one thing, but making a bad job out of your delivery is a completely different matter.

Delivering *fast* in agile (Kanban included) is not about sacrificing the quality of the product. Instead, you should focus on delivering *as fast as possible* while still maintaining the quality of the product at a high standard.

Instead of emphasizing how to be *faster*, emphasize on removing obstacles that might make you slower, such as:

- planning too far out in advance (especially for requirements that haven't been made yet)
- bottlenecks that have not responded to change
- going overboard with your engineering solutions

There is no recklessness in lean. Au contraire, this entire methodology focuses on being fully responsible and fully accountable for your own actions. Instead of rushing a final product to delivery, start with a basic product, deliver it, get your feedback, and then develop incrementally according to the aforementioned feedback.

Empowering the Team

Lean and Kanban are human-centric, meaning that they place the team members at the core of the entire process. This translates into multiple practices and approaches to work, product delivery, and, in the end, into the way your team feels about their work.

Empowering the team means putting them at the center of your methodology. Processes can be changed, products can be adapted to feedback, and budgets can be restricted. But people are rarely actually replaceable, and this is not related solely to the dire situation of the software programmers' job market but to the very dynamics of your team.

Once your team has been shaped, it means you have already invested time and effort into bringing everyone together and working with them. Losing any of the team members is a waste of time, of human resources, and of credibility (in the eyes of the other team members).

If the previously mentioned lean principles are misunderstood, this one tends to be downright forgotten. It is quite easy to understand why people might forget about the power of the humans in a team, especially in highly competitive, fast-paced industries (like the software development ones).

When you focus on delivering fast and on doing a good job, it is very easy to forget about the people who make this happen. Sometimes, it's easy to say *please* and *thank you*. Other times, it's easy to forget that there is life after work (both for you, as the project manager, and for your team members too). And, ultimately, it's easy to forget that adaptability to change is the primordial quality of all agile approaches, so everything can be turned around for a better perspective.

The only exception to this is the *humans* powering your development. When they are burned out, frustrated, and overworked, they will either disengage or call it quits. If you work in software programming, this is especially likely, since a good software developer will most likely find another job in a matter of days or weeks. For you, as the representative of your organization, it might take months before you find the right fit for your team.

Building Integrity In

Like it or not, quality isn't born out of nowhere, and yet, it is one of the core Principles of agile project management in general.

You can do everything right in terms of planning and you can hire the best people, but if you don't build a culture of integrity, self-organization, and responsibility, you risk the very quality of the final product.

Some people tend to take agile methodologies (Kanban included) as too *relaxed*. It's easy to understand why someone looking from the outside might see things as such. Agile teams are frequently pampered with relaxation rooms and PlayStation sessions, table tennis championships, and all the good coffee in the world. Looking at this from the outside, it might seem like the people behind the glass aren't even working.

Beyond all the glam and buzz of working in an agile team (particularly an agile software development team), all of these freedoms and benefits are allowed because there is full trust in each team member's ability to manage their time, their tasks, and their deliveries.

Integrity is a key principle in Kanban (and all agile project management methods, really). And you cannot have integrity if you don't have self-discipline and the genuine desire to *do* better.

Some of the exercises you might want to incorporate in your project management method to promote a spirit of integrity and commitment include the following:

- Pair Programming. As we have mentioned before, Pair Programming can help team members learn from each other and grow as professionals. It can also help them become more self-resilient, more capable of taking responsibility, and more devoted to true honesty.
- Test-Driven Development. Code criteria will help your developers write code that matches the business requirements and objectives. As such, it will help them stay true to a set of basic rules from the very beginning.
- Constant feedback. Again, this has been mentioned before, but we will mention it here as well for a very simple reason: it actually does help people become more responsible and honest about their own work and their skills.
- Maintaining focus. Although this is not an exercise per se, maintaining the focus of your team members on what they actually have to do will help them be more honest about the ways in which they manage their own time. Reduce all context switching and distractions so that your developers and QA engineers can focus on their actual job.
- Automate dull tasks. There's no point in making your developers waste time with copy-pasting and data entry when you can automate these processes. Also, try to

automate pretty much any process that might be prone to human error.

Integrity and devotion aren't easy to build. But once you have set the foundation, they will help your entire project run more smoothly, so that you can deliver faster, better results.

Seeing the Whole

Splitting large projects into smaller iterations and increments is a basic tenet of agile project management in general, and Kanban makes no exception, as has already been touched upon.

However, one of the major risks associated with this type of approach is related to the fact that you can very easily get lost in the myriad of smaller tasks and lose focus of what is actually important.

As a rule of thumb, this is what the Kanban board is for, to allow both you and your team members to take a good look at the big picture every now and again. Furthermore, having a very clear and concise project goal can also help you stay on track and make sure you are all aiming for the same end result.

In software programming, suboptimization becomes a real issue, and it is crucial that you try to avoid it as much as you can. What "suboptimization" refers to is focusing on one or two indicators of success instead of the "whole."

For instance, you might be more tempted to focus on releasing low quality code for the sake of speed. When programmers are more or less forced to *deliver* whatever happens, they might become sloppy, and as such, the quality of their work might drop.

Short term, this might look like a solution because it will allow you to stick to one of the verticals of your plan: timely delivery.

However, doing this will inadvertently have a massive impact on the quality of the product. Sooner or later, the same programmers will have to run through the code one more time to fix all the issues, and this will take a lot more time than if they would have done it right to begin with.

According to the lean and Kanban Principles, you should focus on the entirety of the project. Time, quality, and costs are all part of the "whole," and none of them should be ignored for the sake of the others. In other words, it is always best if you take the time to understand the vicious cycles your team might be prone to, so that you can come up with a better approach, one that will not sacrifice any of the major elements of a successful project.

At the end of the day, optimizing the whole is all about the value stream and how multiple elements come together to deliver a product as close to perfection as possible. Once you have identified the value flow in your team, you will be able to make the right decisions about a multitude of factors that might influence the end result, including:

- how to organize your team
- how to work with colocated and remote teams
- how to cope with inefficient team members

The whole point of this Kanban principle is to ensure that you don't get lost in the details. At the end of the project, your customer won't judge you by the number of *fine* iterations you have delivered, but by the quality of the product, its timely delivery, and how much you stuck to the initial budget.

How is Kanban Different from Scrum?

Items	Scrum	Kanban
Roots	Arose from software development.	Arose from manufacturing domain.
Framework	Within agile framework, which has the core values and principles.	It is actually from Lean production, though now used highly in Agile development.
Roles	There are 3 primary roles in Scrum: 1)Product Owner 2) Development Team 3) Scrum Master	Kanban does not have any specific role. You to start with the existing roles in the organization.
Iterations	Each iteration in scrum is called a sprint. (Usually from 2 weeks to 4 weeks)	There is no concept of iterations in Kanban. It is based on flow based mechanism.
Ceremonies/ Events	Four ceremonies: 1) Sprint planning 2) Daily scrum 3)Sprint review 4)Sprint Retrospective	No defined events. But team can have its planning meetings or retrospective meeting as needed.
Prescriptiveness	Prescriptive, but not a heavily prescriptive process	Less prescriptive, as does not say which meeting to conduct, which roles to do.
Work in Progress (WIP)	In Scrum it is pull and the WIP is per iteration. WIP term is not mentioned in Scrum. But, once product backlog items are committed into the sprint, they should not be changed – hence implicit WIP.	In Kanban, it is pull and the WIP is per workflow state. WIP in Kanban is explicit. The limit is clearly defined and written on top of the columns of the board. Kanban has focus on flow, WIP, batch size and queues.

In many ways, Kanban and Scrum are the ultimate "beginner agile methodologies." They are structured and prescriptive enough to make sense even for someone coming from a highly traditional environment. Yet, they adhere to the agile Principles so well that you simply cannot place them elsewhere.

At their very core, Kanban and Scrum are very similar, so it makes all the sense in the world that they have even given birth to a whole new "offspring" in the world of agile: Scrumban.

The similarities between Scrum and Kanban have already been relayed in the first chapter of this book and so have the differences.

If we have to nail it all down to just a few concepts, keep in mind the following:

- Both Kanban and Scrum are agile methodologies.
- Kanban was born earlier than Scrum and it was initially applied to the manufacturing industry.
- Both Kanban and Scrum use boards for visualization.
- Scrum teams are more structured than Kanban teams.
- Kanban puts more emphasis on waste reduction at all its levels.
- There are no special events or meetings in Kanban (or at least not as prescriptive as in the case of Scrum).

At the end of the day, the main difference between Kanban and Scrum lies in the way they are structured. While Kanban has its own ways of organizing tasks and teams, it doesn't prescribe specific techniques or roles, whereas Scrum does. This might be why Scrum is frequently seen as an easy-to-understand gateway into agile project management and why Kanban is frequently perceived as a "level up."

You are more than free to use a combination of the two methods. For instance, you might find the Scrum board is a bit more explanatory, and you might borrow it into Kanban. Likewise, you might find the Daily Scrum is a useful tool in managing your team. As such, you might want to bring it into your Kanban approach too.

There's no right or wrong as long as you abide by the general agile Principles and stick to the Manifesto. As you will see later on, the lesser-known (but equally valuable) project management methodologies we will present all show similarities to Kanban and Scrum but take things a little further from one point of view or another.

Benefits of Kanban

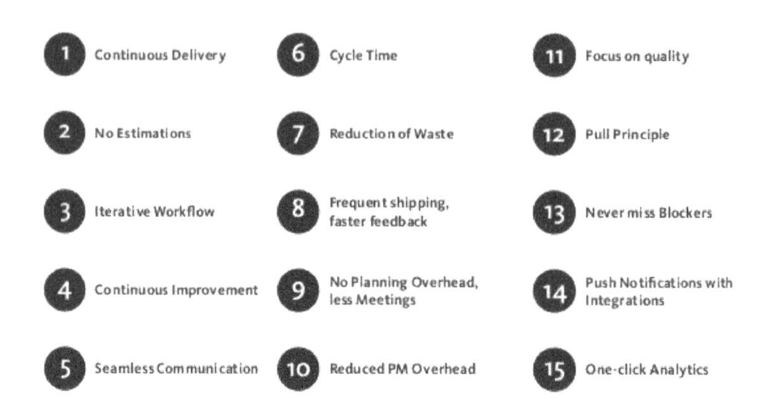

Like all agile project management methodologies, Kanban comes with a pretty generous list of benefits. How each project manager, team, and organization experience these benefits depends from one situation to another. However, some of the most popular advantages of incorporating Kanban in your company/ team include the following:

- Excellent level of flexibility. As mentioned in the previous section, Kanban is far less structured than Scrum. This can be confusing, sure, but it also opens the doors to a world full of opportunities. For once, you can maintain the traditional waterfall organization of the team and projects (and still benefit from all those agile practices). Even more, flexibility tends to be appreciated in modern workplaces not just because it helps deliver better products but also at the team level. With Millennials now being the most numerous segments of the entire workforce (Emmons, 2019), this type of flexibility can earn you loyal, hard-working, talented people in your team.

- Continuous delivery. All agile methodologies focus on delivering on a continuous basis, yes. Kanban makes no exception and actually puts this at the core of the entire approach. As such, this can be considered one of the most preeminent advantages of using Kanban in your organization.
- Waste reduction. No matter how you look at it, waste is one of the major issues of the modern world. We waste paper, food, and time like there's no tomorrow, and while this definitely hurts our future as a race on this planet, it translates into far more immediate and palpable results in the world of business: money. Reducing waste means increased profits, and there's nothing higher management loves more than that!
- Productivity boost. Kanban is almost obsessive about keeping people busy with one task at a time. As such, the productivity of your team will increase. They will stop doing three things at once, and they will leave multitasking at the door when they enter the workplace. Soon enough,

their productivity and efficiency will improve as well. Because, yes, studies show that nobody and *nothing* can actually multitask (Cherry, 2019)!

Obviously, all these benefits eventually transpose into actual business results. Happier and more productive teams deliver better products, they do it on time, they reduce waste in the process, and they tend to stick around for the next project as well. As such, the customer is more likely to be happier as well.

This is not to say that Kanban doesn't show any kind of disadvantages. On the contrary, you should be well aware of the fact that Kanban can lead to low-quality products if, for example, the Kanban board is outdated. Furthermore, Kanban tends to be less time oriented because it doesn't impose actual timeframes, and as such, this can lead to problems as well.

All in all, though, used right, and used in combination with other project management systems, Kanban can make your production process more fluent and more efficient across its verticals.

Kanban and VersionOne

As we were saying earlier, both Kanban and other agile project management methods can benefit from using the right tools. These days, agile project management tools are recognized at their true value, so much so that you won't see many agile teams without at least one of the major tools (which we have already discussed in the first installment of this book series).

VersionOne is one of the best tools to use when you want to implement Kanban because its very structure is built to be fully congruent to the basic Principles behind this specific agile approach. Although VersionOne can be used with pretty much every other agile method out there, it does provide a ready-made Kanban board that will help you implement the methodology in an easier, smoother way.

Some of the other features that will help you run your agile projects better include:

- the ability to connect distributed teams
- the ability to connect teams that use different agile methodologies (including Scrum and XP)
- the ability to automate project tracking
- the ability to collect data about the performance of your team and the efficiency of the entire project development process

We suggest you try VersionOne's free version and see how it works for you. Of course, you can use any of the other tools we have already mentioned in the first part of this book series, but if you are looking for something that is a little more Kanban inclined, this specific tool might be what you are looking for.

Kanban can be a real steppingstone for your team, and it can completely change not only how you work but how you perceive the entire concept of work itself. Although it might sound strict, Kanban is one of the more relaxed "entry agile" methodologies out there, one that will allow you to play around with team structures and agile concepts as you see fit for your specific needs, for your organization, and for your team's mental structure.

Implementing Kanban might not be 100% easy, but it definitely is one of the least headache-inducing agile methodologies to introduce to a team that has been working on a waterfall framework until now. Because it leaves plenty of room for documentation, charts, and other traditional project management tools, Kanban is a relatively good fit for larger organizations and corporations as well (and if you are ever in doubt, check out Toyota and how they use lean manufacturing and Kanban Principles in their work).

At the end of the day, no agile project management methodology can be a saving grace if you don't give it a real chance, and Kanban makes no exception. Study it, implement it wisely, and watch it bring your profitability, proficiency, efficiency, and overall productivity to a whole new level.

Our next chapter is dedicated to a more intense agile project management methodology: XP (Extreme Programming). The name shouldn't scare you, though. Same as Scrum and Kanban,

XP abides to the same Principles of agile project management; it just takes it up a notch in terms of just *how* in-depth it goes with said Principles.

We definitely advise you to look into XP and the other project management methodologies we will describe further on in the book. They might not be as wildly popular as Scrum and Kanban, but they might be just what you are looking for!

Chapter 3: Extreme Programming (XP)

Extreme Programming (also known as "XP") was one of the first agile project management methods developed almost exclusively for the world of software development. Same as Scrum, XP came as a solution to problems that were specific to the 1990s (and which had been escalating since the 1980s, actually).

The fact that XP is more commonly used in software development environments makes it less known outside of the world of coding but no less valuable. The *extreme* nature of this methodology has somewhat washed out over the past two decades since its incipient stages, but the basic principles stay the same.

Indeed, XP seems to be a less popular choice in and outside the world of software development. Studies show that, in 2015 (Ropa, 2015), less than 1% of the respondents were using XP as their core agile methodology.

We believe it is still relevant, though. Even if you don't embrace it altogether and take the extreme path it suggests, this methodology can still provide you with valuable tools and practices, and many of them can be successfully incorporated into other agile approaches (including Scrum and Kanban, as it has been shown over the past chapter).

To understand Extreme Programming, you must first understand why there was a need for such a swift turn in practices. The origin of XP can be traced back to two main issues software development houses in the 1990s were facing:

- the need to adapt to an increasingly fast-paced world (consider the fact that the 1990s represented the rise to power of the dotcom bubble)
- the need to focus on object-oriented programming, rather than procedural programming

At its very core, XP is much more technical than other project management methods, and this might be one of the reasons it hasn't transcended the borders of the industry in which it was born (like Kanban and Scrum did, for example).

Extreme Programming relies on discipline even more than its agile "sisters." Its main goal is similar to those of every other agile methodology in the world: to deliver a better product in a shorter amount of time.

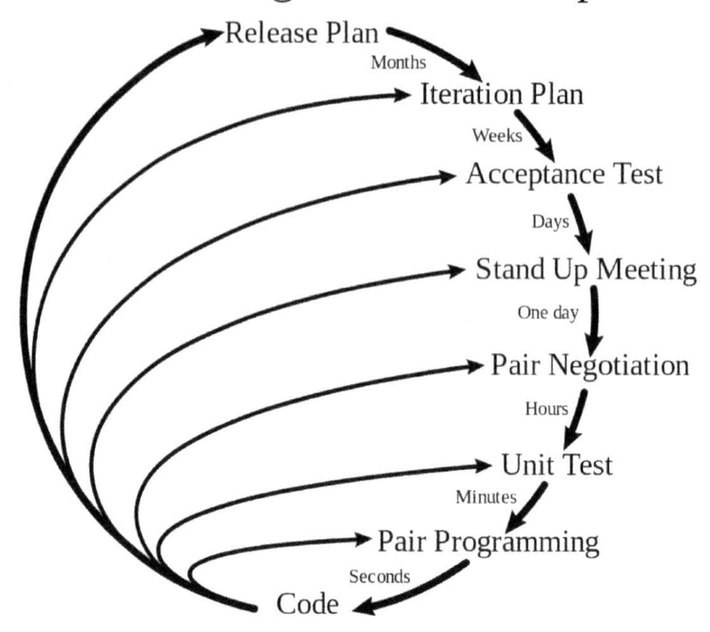

To reach this goal, XP relies on four main activities:

- Coding. In the Extreme Programming paradigm, this is the absolute key to developing quality products. Coding is King!
- Testing. While XP argues that codes are the most important element of the software development process, it also admits that testing *needs* to be done to ensure written codes are correct. There are three types of testing to be done: unit tests (to verify that each feature is functional),

acceptance tests (to verify that the user requirements have been properly understood by the developers), and integration tests (to verify how different sections of a software connect to each other).

- Listening. Constant feedback and discussions with customers are encouraged so that developers understand the requirements. Moreover, developers are also encouraged to explain why some requirements may or may not be possible (and how they are or are not). Of course, since in most cases it is unlikely that the customer will be as technical as the developers, these explanations have to be provided in a business tone, rather than a technical one.
- Designing. In an ideal world, coding, testing, and listening should be *just enough*. However, XP recognizes that most software systems are complex enough to need proper design that connects the different features and makes the program easier to understand and use.

Aside from the core actions of Extreme Programming, it is also worth mentioning that this methodology has embraced the agile Principles and reshaped them in a form that is more congruent to the basics behind its specificities. As such, the values promoted by XP project management are:

- Communication. Developers have to constantly communicate among themselves, as well as with their customers.

- Simplicity. Everything should be kept as simple as possible, from the code itself to the processes behind the development.
- Feedback. As this is a basic tenet of agile project management, it hasn't been missed from this specific methodology either. Feedback is always encouraged and it should be followed at all times.
- Courage. In XP, courage is manifested in many ways. For once, the team must have the courage to program for today, rather than tomorrow (which avoids long and winding design processes). Furthermore, they should know when to delete code and when they should leave it as it is. Last but not least, programmers should have the courage to accept that sometimes a problem might not be easy to solve but that persistence will eventually help.
- Respect. This value is manifested both at the level of the individual (team members should respect themselves and their work, and as such, they shouldn't make compromises) and at the level of the team (everyone should respect each other's work).

In essence, XP is not much different than other agile project management methods. What makes it different, however, are the specific practices it employs. We will dedicate the remainder of this chapter to exploring these specificities. As mentioned before, you can borrow all of them into your project management approach, or you can nitpick them and settle on those which prove to be valuable for your team's needs.

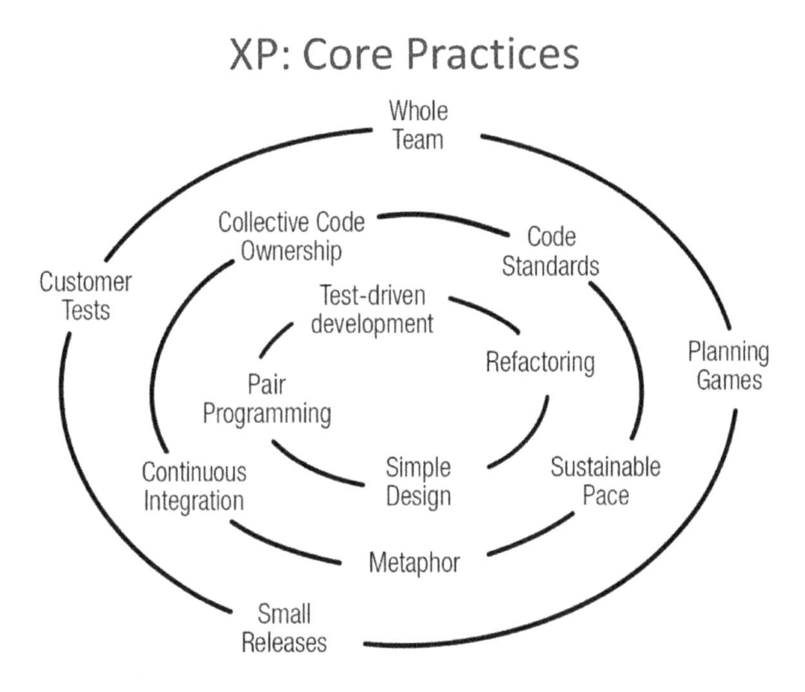

Planning Game

What is known in Extreme Programming as the "Planning Game" is nothing more than the specific way in which XP sees the planning of a project.

What is very specific about this planning method is the fact that the customer is always involved in the process. In short, the customer will bring forward all the information they have about the value of their requested product, while the development team will bring forward the information related to the cost of the product development.

The dynamics between the customer and the development team are meant to achieve one main goal: to reduce the time it takes to develop the product by removing intermediaries and helping the customer communicate directly to the programmers.

As you can see, this is quite a difference between XP and Scrum, as Scrum has at least two intermediaries between the customer and the development team (the Product Owner and the Scrum Master).

Although this is not exactly set in stone, most XP project Planning Games include four steps:

1. Create and select the story. If there is no story created, the customer must create one. If the stories have already been created, then the customer should simply pick one from the list.
2. Estimate the story. Once the customer has picked the story, the ball is in the developers' court. In other words, at this stage, the programmers should estimate how much time it will take for the story to be delivered and how much it will cost.
3. Repeat the process with all the stories.
4. Prioritize the stories. Once all the stories have been estimated, they will be prioritized by the customer

according to the requirements they want the final product to meet.

You will encounter similar procedures and games in other agile project management approaches as well (including Scrum, for example). The main difference lies in the involvement of the customer in the entire planning process (a feature XP puts a lot of emphasis on).

Small Releases

Splitting the project into smaller increments (stories) is very common in agile project management. The main purpose of doing this is twofold. On the one hand, it allows teams and project managers (where they are present) to better manage the workload. On the other hand, it allows teams and project managers to deliver better and faster.

Small releases are an integral part of Extreme Programming as well and, same as with other agile methodologies, their main goal is to allow for better control in terms of timely delivery, quality, and adaptability to change.

In XP, small releases translate into actual mini-product releases that are made available to the customer. Once feedback and data are gathered on what works and what doesn't, the product will be improved throughout the following iterations until it is ready to be published in its full format.

To make sure your small releases achieve their main goal, you should keep in mind to:

- Be ready. Just because this is an agile approach, it doesn't mean that you shouldn't plan ahead. In XP, you are required to know what each iteration/ story will deal with, as well as how long each iteration will take to be completed.
- Communicate. You should *always* make sure to communicate with the customer, as well as among yourselves, as a team. All the features should be discussed, both in terms of how-to time and budget them and in terms of how to actually code and execute them.
- Feedback implementation. As we have also mentioned before, there is no point in asking for feedback if you don't take action accordingly. Continuous improvement is crucial in all agile project management methodologies!
- Experiment. You shouldn't be afraid to experiment with new methods, new codes, and new solutions. At the very worst, they will not work. At the very best, you might have just found a new way of solving a problem that is most likely older than your project.

Small releases can be extremely helpful when you want to make sure you have granular control over what happens in the project and when you want to make sure the end product is fully satisfactory for the end user.

Customer Acceptance Tests

We have already touched upon the concept, but we would also like to expand on it a little more, since it's a practice quite specific to Extreme Programming (one that can be incorporated in other agile approaches too).

Customer acceptance tests are quick tests (frequently automated tests) that involve the customer in a measure that is equal to their involvement in the Planning Game. Basically, what the customer will do is translate each story into an acceptance test. When the story is delivered, the customer will specify different scenarios to see if the product functions as they wanted it to function. If it doesn't, the product will go back into development, and the feedback will be incorporated in the new iteration.

Before the product reaches the customer's hands, it is quite important to ensure that it has been properly run through the Quality Assurance team's tests. This also means that programmers and QA engineers should have a pretty close relationship and that they should communicate with each other quite efficiently (especially given that there is no project manager here to mediate the relationship between the two).

No user story should be considered as "Done" until it has been approved by the customer. One story can also run one or more customer acceptance tests, depending on the complexity of the product and how good it is when it is released the first time.

Simple Design

In agile, everything should be maintained at a very high level of simplicity. Everything should be easy to work with, easy to implement, and easy to use when it comes to the end user.

Simple design is one of the core values of Extreme Programming because it allows the very architecture and presentation of the product to stay clean and tidy. At the same time, it is essential to mention that "simple design" is not one and the same with "simplistic design."

You want complex programs to be easy to use. You don't want to ultrasimplify the complexities of your program. To ensure this happens, keep in mind to:

- Do your research and be aware of any potential flaws in your design.
- Code simply; do not overcomplicate it unless it is absolutely necessary.
- Don't be afraid to postpone decisions to a moment where you can have full insight into the data your business is collecting.
- Document all the design choices and changes in your backlog, so that you can go back to them as a reference for the future.

Simple design can help you make the development process easier for your customers, and it can help you deliver the kind of product they can easily use. While your first story might not be exactly what the customer is asking for, these feedback cycles are actually helpful because they will help you understand everything better.

Don't look at simple design as an impediment on your way to success. It is there to help!

Pair Programming

Pair Programming is a concept we have already introduced in our previous chapter, one that has been borrowed not only by Kanban (as we have shown) but by other agile methodologies as well.

Pair Programming brings two developers together. One is the Driver (the person who writes the code) and the other one is the Observer (the person sitting in the back, verifying the code written by the Driver). Ideally, both the Driver and the Observer will change places and one will become the other, so that they can both have the same amount of time in front of the keyboard.

Moreover, it is recommended that both the Driver and the Observer have the same amount of experience under their belts. This way, neither one of them will feel left behind or "pushed" in any way.

The Pair Programming practice relies on the four-eyes principle applied in multiple industries. For instance, in book publishing and translations, it is believed that two pairs of eyes are more likely to spot a variety of mistakes (as opposed to having just one pair of eyes check the work). Furthermore, in business, you will frequently notice that documents have to be signed by two people (e.g., the CEO and the CFO). This too is a permutation of the four-eye principle.

In XP and the other methodologies that have employed Pair Programming, the four-eye principle relies on the idea that two people are more likely to write clean code. Both the Driver and the Observer have to work together by coding and testing the features they are building, and, in the process, both of them are more likely to spot mistakes along the way.

Pair Programming brings with it a long list of benefits, such as boosting collaboration and communication, improving the overall quality of the code, and delivering in a shorter amount of time.

One of the main drawbacks associated with Pair Programming is that simply having the hard skills (the coding knowledge) is frequently not enough for this to work. In order for two people to work as closely as this, they need to also possess soft skills (like the ability to communicate properly and a general ability to work as a team, for example). Without the soft skills, it is easy to see how this practice can very quickly turn into a bad joke at the very best or a complete disaster at the very worst.

Test-Driven Development

The concept of Test-Driven Development is a very good fit for the world of agile in general, and even more so when it comes to Extreme Programming, where things tend to be even more on fast forward.

Test-Driven Development is a type of software development process that focuses on a very specific, very short development cycle:

- User requirements are turned into test cases.
- The tests are run for verification.
- The coding is done.
- The tests are applied on the code.
- The code is cleaned during the refactoring phase.
- Everything is repeated until the final product is delivered.

Test-Driven Development shows one major benefit: it helps programmers stick to cleaner codes, as opposed to getting lost in intricacies of any kind. Principles such as *Keep It Simple, Stupid* and *You Aren't Gonna Need It* are commonly applied to Test-Driven Development precisely because it allows the developers to focus on clean code that is well designed and simple to use, rather than overdoing it and creating a code that will be more difficult to change or clean up.

There are some practices you should avoid in Test-Driven Development as well:

- Your test cases should not depend on the previously executed test case systems. Instead, you should start a unit test from a preconfigured state at all times. Likewise, interdependent tests should not be run either, as they can lead to false negatives and fail the entire process.
- You shouldn't go overboard with your test cases either. For instance, trying to create a test that will inspect everything under the Sun means that you will make your entire test more prone to mistakes, and that defeats its purpose.
- Your tests shouldn't be slow running either. Since this is a very fast-paced environment, both your coding and your testing have to move at the same speed as the project. Of course, this is not to say that you should sacrifice the quality of the code, but you shouldn't linger too much on any of the steps involved in the development process.

Test-Driven Development may not be all about advantages, for certain. But in some cases, using this technique might actually make the difference between a project delivered on time and one that is late.

Refactoring

This is one of those concepts that sounds much more difficult than what it actually is. Commonly used in XP and other agile methodologies, Refactoring is a practice that focuses on simplifying the code to the point where changes and bug fixes can be very easily implemented.

At the same time, it is very important to mention that Refactoring *does not* mean that you should sacrifice the functionality of the code in any way.

To ensure the efficiency of the Refactoring process, it is of the utmost importance to be constant in practicing it. In other words, Refactoring should be a regular affair. You can do it daily or weekly, but it needs to be done with regularity, every single time. Moreover, it should follow the same simple process every time as well. For the team, Refactoring should become like a second nature, a routine they implement into their work in the same manner they grab a cup of coffee every morning when they get to the office.

The Refactoring process can be adapted according to needs, but most often it will follow the same pattern:

- Find or create a test for the part of the code you want to refactor. This will help you make sure you do not ruin the integrity of the code in the Refactoring process.
- Run code simplification and improvement as you see fit.
- Apply the test on the unit you have modified.

- Repeat until you have Refactored all the units that need to be changed.

The Refactoring practice is simple and it can be a real lifesaver. Say, for example, that your customer wants to change something about the product when it has already been developed. The simpler the code behind it is, the easier it will be for the team to dig in and apply the needed change.

It's a win-win situation for everyone: the customer is happy with the speed of execution and with the overall quality of the product, while the team is happy that they don't have to navigate ridiculously complex codes to determine exactly where the change should occur (and how).

Continuous Integration

Continuous integration is not a concept that is exclusive to Extreme Programming in general. Rather, it is a concept that keeps popping up in the entire world of agile project management. However, it is more common to see this practice applied in XP and the other more extreme agile methodologies precisely because it relies a lot on automation and on fast-speed development processes, which makes it more suitable for these specific agile approaches.

Continuous integration is a practice that works especially well in agile project management precisely because agile focuses on developing products in an incremental way. With the help of continuous integration, the different stages of development are more easily integrated into the whole without too much time invested in the integration process and without the potential errors that might ensue.

In short, continuous integration refers to keeping the code in a central repository where developers and testers can easily access it. When the system is ready, the code can be tested right away, using the aforementioned repository. This allows the team to run tests as they move along with the code, thus minimizing the time needed to deliver working software.

These days, continuous integration is even easier to perform than what it used to be like back when Extreme Programming was first shaping up. This is mostly due to automated systems that allow programmers to easily move along with the code and test it at the same time. Machine learning has made everything easier, allowing for better productivity and efficiency in software development (and not only these areas!).

Collective Code Ownership

As we were mentioning before, Extreme Programming is an agile methodology that is quite extreme in the level of flexibility it allows. The development of an XP project is *so* flexible that everyone is encouraged to constantly bring improvements to any part of the project that they might want to.

The code is collectively owned by everyone in the team or, in the case of open-source projects, by everyone who wants to chip in and simply make the code better from one point of view or another.

The XP team is completely decentralized and destructured, and this might be a concept pretty difficult to grasp if you come from a very traditional environment. In fact, it can be pretty difficult to grasp wherever you may come from, including the more structured agile methodologies out there (such as Scrum and Kanban, for example).

In an XP project, everyone is responsible for the entirety of the code. This means that the team doesn't rely on a chief architect to have all the good answers, all the time. Instead, the team relies on its entirety to make the software a working product.

Extreme Programming understands, perhaps better than many other project management methodologies, that nobody is perfect, not even a super-experienced chief architect. Instead of placing the responsibility for the final product on one or two people, XP shares the load. As such, the code is collectively owned, and everyone is invited (and encouraged) to participate in it.

This might sound like chaos, but years and years of experience have proven that, yes, this is a more reliable way of treating software development. With continuous integration being such a massive part of Extreme Programming projects, the amends made by the team to any part of the code are hardly even noticed. Instead, everything simply flows smoothly.

Collective code ownership also means that you have to place complete trust in your team and their skills. Without this trust, it's easy to see how you might become suspicious of your team members and how you might not be able to see their improvements as actually beneficial.

On the upside, this level of trust in a team can also empower them and make them more responsible at an individual level. It might be surprising, but most of the times, people don't disappoint when so much trust and responsibility is placed in their hands. Rather, they constantly double- and triple-check all of their work to ensure that their codes and their amends will not have to be cleaned up by the rest of the team.

Coding Standards

One of the major downfalls of practices like the Collective Code Ownership is related to the fact that, well, different people can see a feature developed in different ways.

Programming is very much like every other language: you can use it as you see fit and still get the message across at a basic level. As such, you need standards to ensure that people in the same context will *speak* the same language, not just in terms of basic linguistics but also in terms of tone, style, and voice.

The same principle applies to XP projects as well. Coding standards have to be set in place to ensure the consistency of the code across the entire project. Otherwise, you might soon learn that there is a major issue with so many people being responsible for the entirety of the code: they think differently, and they might not be on the same page in terms of how they organize their code, how they relay the information to the computer, or even how they understand the concept of "Done."

Although there is no prescription for this, most often, there are three main categories of coding standards:

- what is mandatory (the rules that have to be followed at all times)
- what is considered to be a good practice
- what is considered to be a recommendation (tips that should be followed, but they are not mandatory, and the developer should use their better judgment in deciding whether or not to use them).

Furthermore, there are different types of coding standards as well:

- formatting (such as, for example, how white spaces are used)
- code structure (such as, for example, project layout, classes, resources, and so on)

- naming conventions (how the team names their classes, methods, and so on)
- error handling (how objects in the code should handle errors, reporting, and logging, for example)
- comments (how to explain the logic of the code)

Coding standards are quite important when you want to minimize the risk of a messy code (and thus, a code that is more likely to have bugs and less likely to be easily adapted to the changing requirements of the customer).

Metaphor

The System Metaphor (or more simply put, the *Metaphor*) is, at its very core, nothing but the communication that backs up the development of a product. The Metaphor is a truly essential part of an XP project (and it is encountered, in different forms, in other agile project management methodologies as well). At the same time, the System Metaphor is also very poorly understood (and commonly forgotten altogether).

Basically, what the Metaphor does is communicate with the customer and the stakeholders in their own language. It is a translation of the technicalities and the jargon that are used within the team for people who are involved in the project but aren't exactly "technical."

There are two levels at which the Metaphor happens:

- A unified story: a story that hovers over the entirety of the project and must be used by all those involved, at all times.
- A shared vision among the different participants in the project. No matter where they come from, everyone involved in this project will have a shared vision (coordinated by the unified story, of course).

The way you use Metaphor in your project can actually make the difference between a good understanding of the product requirements and a bad one, between customers who do understand your processes and customers who don't, and, in the end, between products that satisfy the needs of the clients and products that don't.

Although it might be a less technical part of the XP project, the Metaphor is as important as any other practice associated with this agile methodology. Use it wisely and it will help you deliver better products!

Sustainable Pace

Sustainability and scalability have been frequently touted as the two major downfalls of agile project management in general.

The sustainability of XP has been especially problematic and argued over the decades, and the debate that ensued lies at the basis of why Extreme Programming isn't as popular these days as it used to be (or at least not in its purest form).

Due to the fast-paced environment of the software development world in general, many XP teams used to feel the need to go at a speed that was much higher than the average. As a result, burnout and mistakes happened as well, proving that going well overboard with what you can do is never a long-term plan.

Contrary to what many people would be tempted to believe, XP does not advocate for a lack of sustainability. On the contrary, proper Extreme Programming should be done at a pace that is sustainable in the long run.

There are multiple reasons for this. The first one is related to the fact that it is simply impossible to expect a very high speed of delivery and a very high quality of product at the same time or at least not long term. It might happen for one sprint or two, but it cannot be a sustainable strategy for the duration of an entire project, much less a project that will be finished in a matter of *years*.

Furthermore, Extreme Programming is highly reliant on team members' abilities to actually cooperate and communicate with each other. Unfortunately, that rarely happens in teams that hardly know each other due to high job mobility. Unfortunately, as well, job mobility and unstable teams are two of the consequences of burnout.

As such, it is highly important to make sure you move at a pace your team feels comfortable with. It may not lead to super-fast delivery times, but it will maintain the integrity of the code, and maybe even more importantly, it will maintain the integrity of the team.

Extreme Programming has been dubbed as one of the least scalable agile methods, and we are sorry to say it, but we tend to agree with this statement. Because it relies so much on the cohesion of the team, XP cannot be scaled to very large teams and corporations.

What *can* be done, however, is integrating XP-specific practices (such as the Paired Programming one) into other agile and/or traditional methodologies. Doing this will allow you to get the best of the two worlds: the high speed and high quality that XP projects are frequently associated with *and* the long-term sustainability of applying such methodologies to projects and organizations that go beyond your average XP team size.

At the end of the day, it can be quite easy to understand why XP has decreased in popularity. Instead of remaining closed within the limitations of each methodology, agile has proven, time and again, that flexibility is its main tenet, and, as such, it has grouped together practices pertaining to different methodologies to create the perfect environment for each situation.

Scrum, Kanban, XP—none of the major agile project management methodologies are used in their pure forms, most of the time. Instead, companies and project managers rely on a combination of practices that are fit for their teams and their specific situations. We encourage you to test out Extreme Programming practices too. They might prove more than useful!

Chapter 4: The Crystal Method

While you may have heard about Scrum, Kanban, and even Extreme Programming, the Crystal Method might be more of a mystery to you if you are new to the agile project management world.

Indeed, the mainstream agile methodologies have migrated past the borders of their agile world, and they are now common even in companies that deem themselves as adhering to a traditional waterfall approach.

When it comes to the less popular methods, however, things get a little more complicated. Methods like the Crystal one is incredibly powerful and useful for a wide range of situations, but they tend to be less prescriptive and thus less adaptable to teams that are new to agile. As a result, the Crystal Method has not yet made the leap into the mainstream agile project management framework.

We want this book to be a comprehensive overview of the main agile methodologies employed by worldwide organizations. As such, we have decided that we will include the less popular, but still extremely valuable, methodologies here as well. Crystal falls in this category because, as you will see later on, it can provide you with a lot of benefits.

We do not claim that this chapter covers everything Crystal is about. As mentioned in the introduction of this book, we do not claim that the entire book encompasses everything all these agile methodologies are. Instead, our purpose with this specific chapter is to introduce you to the Crystal Method and allow you to make your own choices. You might be surprised to find that Crystal makes a lot of sense, and you might decide that it is the right direction for your team and organization too.

Even if you don't choose Crystal to guide you through the intricacies of the agile world, we still believe there is value in learning about it. The more you know, the more in depth you can go with your agile approach and create a "program" that allows you and your team to perform at your maximum capacity!

What Is the Crystal Method?

This is a lesser-known fact, but the Crystal Method (or the Crystal *Methods*, as the approach is frequently referred to) was born in one of the biggest and most resilient IT companies in the world: IBM.

It sounds odd to think that one of the least known agile methodologies out there was born in one of the most famous IT businesses, but the low-prescriptive, very lightweight nature of Crystal explains why it isn't one of the more popular choices for companies these days.

In addition to this, Crystal is one of the first agile methodologies ever developed, at the beginning of the 1990s. Come to think of it, there was a one-decade gap between the moment Crystal was developed and the first official Agile Principles were laid down in Utah, so it makes quite a lot of sense that the Crystal collection of practices has been left behind in the fog of time.

This doesn't make Crystal any less valuable. On the contrary, we believe that it can be a very good approach, especially for teams that are already at least somewhat used to agile project management.

Sometimes, it takes going way back to the roots to discover the true essence of something that has evolved over the years, and agile project management is no exception to this general statement. The Crystal Method might be exactly what you need to rediscover the true roots of agile and why it was such a groundbreaking, earth-shattering framework when it first started to trespass the closed world of high-end software project management.

The Crystal Method is very simple in essence, and yet, applying it with no limitations whatsoever can be quite tricky, precisely because it leaves the door open to a world of opportunities in terms of the specific practices you may or may not want to employ with your team.

The complications of Crystal arise when you try to understand the fact that its value lies in the lack of prescriptiveness and the ways in which it can be adapted to teams that are wildly different in nature.

In other words, Crystal is only as hard as you make it. Beyond the basics (which we will relay in the following sections of this chapter), Crystal is nothing more than one of the incipient agile methodologies. It relies on the same basic Principles; it's just that they are less polished, less prescriptive, and less limited at the same time too.

If we had to summarize the Crystal Method, we would simply say that it is a collection of primordial agile practices adapted to the specificities of various types of teams. Although incipient and apparently dated, these practices continue to make sense in today's environment, particularly as more and more companies are looking to scale up the agile methodologies they have been using to date.

How Does Crystal Operate

When he first analyzed the dynamics of teams at IBM, Allistair Cockburn noticed that there are two main tenets that make them more efficient:

- the ability to streamline and optimize their own processes according to the workload

- the adaptability they have when it comes to the uniqueness and specificities of each project

Later on, these two concepts became the foundation of the Crystal Method in a way similar to how the Agile Manifesto rules over the entire agile framework.

While he was developing the Crystal Method, Cockburn also noticed that there should be a clear differentiator between methodology, technique, and policy. This helped IBM understand where Crystal begins and where it ends, and it continued to help organizations everywhere delineate between what their project management approach is and what other regulations of their businesses are.

In short:

- A methodology is a set of elements (such as practices and tools, for example).
- A technique is related to skill areas.
- A policy is related to what *must* be done in an organization.

Furthermore, Cockburn also stated that the new set of methods he was developing would be focused on six main areas:

- the people
- the interactions between the people
- the sense of community people gets at work
- the skills people have

- the talents people have
- better and more efficient communication between the people

As you can see, the Crystal Method is a clearly human-centric view on project management. By far and large, what Cockburn developed back at the beginning of the 1990s might have set the foundation for the first point of the Agile Manifesto later on, in 2001.

Same as the Agile Manifesto and its adjacent Principles, Crystal says that processes are never more important than the people. In fact, according to Cockburn, people should come first in all circumstances, and processes should come second. The *team*, as Cockburn puts it, is the core of the project because they are the ones holding the skills and the talent, and as long as these are abundant, processes follow after.

Additionally, Cockburn also identified four main behaviors people have in relation to work:

- People are meant to communicate and crave for proper communication. It is worth mentioning that this is a concept all agile methodologies support. But it is equally important to mention that this concept was perceived differently back in the 1990s and the beginning of the 2000s, when online communication had not yet reached the level of accuracy and real-timeness of today's technology.

- One of the main issues people have is being consistent over the course of a longer period of time. As such, the project management approach should be built in a way that keeps the team continuously engaged and supports software development in a sustainable way.
- People are not constant either. They have good days and bad days, they may or may not perform at their best in one place or another, and they will always be the same. As such, the processes behind the project management approach should adapt to this type of change as well.
- People are inherently good. They want to be good citizens, they want to take care of their peers, and they want to do a good job. If you enable them to be their best, they will be, and as such, they will perform better on their day-to-day duties at work.

Although this might sound odd coming from one of the largest corporations in the world, the Crystal Method is one of the easiest ones to adopt in terms of the rules it brings forward precisely because it has designed mini-frameworks for a variety of situations.

More specifically, Cockburn split the entire Crystal Method into several categories, organized according to five major criteria:

- the size of the team/ project
- comfort
- discretionary money
- essential money
- life

The first criterion is pretty straightforward: the size of the team influences how it should be managed.

The other four, however, are related to the impact the system could have on the mentioned verticals if it doesn't work:

- how it will impact comfort
- how it will impact the disposable income of the user
- how it will impact the essential income of the user
- how it will impact the life of the user

According to these five criteria, Cockburn has categorized the Crystal family of Methods into five color-coded groups, as follows:

1. Clear: for teams of up to 6-8 people, low to no impact across all verticals
2. Yellow: for teams of up to 20 people, low to medium impact across all verticals
3. Orange: for teams of up to 40 people, medium to high impact across all verticals

4. Red: for teams of up to 80 people, high impact across all verticals
5. Maroon: for teams of up to 200 people, very high impact across all verticals

Aside from these five basic categories, you might also encounter adjacent Crystal Methods used in specific contexts:

- Crystal Orange Web (used for web products)
- Crystal Sapphire and Crystal Diamond (used for large-scaled projects that can have a dangerous impact on human life, like aircraft software, for example)

It is also essential to note that there is quite a lot of flexibility when it comes to how each category applies to your team and project. However, should the team grow over time, it is recommended to upgrade to the next color-coded category, rather than continue to apply the same methods and try to scale them up.

This might all sound very prescriptive, but it is quite the opposite in most respects. At the same time, keep in mind that this methodology was not born in a small software development house. It was born in a company that was already a mammoth corporation by the time Crystal came into play. As such, you should see the lack of prescription through the prism of a company that was very well-grounded in traditional approaches too.

One of the areas in which this is more apparent is the way in which Crystal Methods assign specific roles to team members and stakeholders. While you will encounter a pretty deep level of role-defining in Scrum as well, the very nature and the names of the roles assigned in a Crystal project will most likely be more familiar to those coming from waterfall approaches.

Some of the roles you will encounter in a Crystal project include the following:

- Project Sponsor (customer or internal stakeholder)
- Senior Designer/Programmer (the equivalent of team leads)
- Designers/Programmers (Business Class Designers, Programmers, Documenters, etc.)
- Architect
- Requirements Gatherer
- Coordinator
- Business Expert
- Project Manager
- Design Mentor
- Lead Design

... And so on.

We will not dwell too much on the specific differences between the different roles assigned in a Crystal project. For the most part, they are quite self-explanatory and quite similar to traditional roles you will encounter in many other software programming projects.

Furthermore, we will not dwell too much on the differences between the different Crystal family members either. Each of the colors is assigned to a type of project. Different specific methods are meant to be employed according to the criteria that were already tackled earlier in this section.

If you want to take a specific Crystal approach, we highly encourage you to go in depth with your knowledge on that specific family member. Same as in the case of XP, even if you don't end up using that particular approach, you might still find practices your team is congruent with.

Crystal Method Characteristics

The basic characteristics of the Crystal family of Methods might be easy to understand from the introduction we have made in the previous section. However, for the purpose of clarity, we feel the need to also relay the specific characteristics that bring all Crystal Methods together.

Things are quite simple in this sector. There are three main characteristics the Crystal Methods have in common:

1. They are all human-powered. As it was shown in the previous section, manpower lies at the very core of the Crystal methodology. People are the blood and veins

running through projects and making them come to life, and, as such, all Crystal Methods (regardless of the color they were assigned) put massive emphasis on the importance of the human resources.

2. They are ultralightweight. All agile methodologies are lightweight, but methods like Crystal take this up a notch and become *ultra*lightweight. What this means is not that team members are allowed to do whatever they please, but that documentation and reporting are less of a point of focus in Crystal than they are in other methods. This is also why implementing Crystal from a waterfall standpoint might not be as easy as it is generally believed by those who go through its basic principles and M.O.
3. They are adaptive. Like all agile methods, Crystal methods are adaptive. They embrace change as a natural part of the process, and they embrace the differences between the different teams as an equally natural element. As such, Crystal Methods are adaptive, flexible, and interchangeable down to their very core.

That's it. These are the characteristics that bring all the Crystal Methods together and tie them to the larger world of agile project management.

Properties of the Crystal Method

Aside from the basic characteristics shown in the previous section, the Crystal Method has a very stable set of properties as well. As you will see, these properties are similar to the Agile Principles. So, for the most part, they will most likely not surprise you.

The seven main properties of the Crystal Method are as follows:

1. Frequent delivery. Same as all the other agile projects, Crystal projects should be delivered in smaller increments, rather than larger chunks (or entire products at the end of the project). This will help you ensure that you don't put a lot of money and energy into a product that will not be well received (by the customer or by the end market).
2. Reflective improvement. Constant and continuous feedback lies at the foundation of agile in general. So, it makes all the sense in the world that Crystal would adopt this principle as well. The more feedback you get and the more you reflect on how to improve your product, the better the end result will be.
3. Osmotic communication. This might sound very fancy, but it refers to a very high level of communication among team members. For smooth projects, you need to constantly encourage people to talk to each other and communicate their issues, opinions, and suggestions.

4. Personal safety. As mentioned before, Crystal is one of the most human-centric approaches you will ever encounter. As such, it puts a lot of emphasis on personal safety as well. This is not so much related to personal safety in the sense of ensuring a lack of life-threatening conditions at work (which goes without saying, right?), but in the sense that people should be encouraged to speak up. Yes, this means they should speak up even when they don't agree with the majority.
5. Focus. This property is twofold. On the one hand, it speaks about each team member's ability to focus on one task at hand and how to deliver it better. On the other hand, it speaks about the focus of the entire project, defining clear objectives and goals and ensuring the entire team is on the same page in all respects.
6. Easy access to expert users. In Crystal projects, you are encouraged to always get feedback from the end users. Focus groups and surveys might help with this, but do keep in mind that you might need interdepartmental cooperation for this.
7. Technical environment focused on test automation, configuration management, and frequent integration. The more automated your processes are and the better your management and tracking tools are, the more likely it is that your team's skills and talents will be focused in the

right direction. As such, the final product will be better from all points of view.

Why Is the Crystal Method Useful?

All agile project management methodologies are useful in their own way. The Crystal Method makes no exception in this area either.

Overall, the Crystal Method is considered to be useful for the following reasons:

- It's fairly easy to implement, both due to the high level of categorization it includes and to the high level of flexibility it allows.
- It's easy to implement its specific practices with other agile and nonagile methodologies.
- It focuses on people and communication, and in a world that seems continuously disconnected from its human nature, this is a *big* advantage.
- It allows continuous integration and incremental deliveries, which helps deliver better products in the end.
- The processes are configurable, and you don't have to worry about documentation too much.
- It actively involves the end user, and this means that you will get better, more accurate feedback that suits the actual market.

The lack of specific guidelines can be quite confusing when it comes to the Crystal Method, but it is more than worth mentioning that Crystal was not meant to become a policy or a set of actual techniques from the very beginning (this is specifically why Cockburn made the distinction between the three terms).

At the end of the day, Crystal can provide you with a mindset at the very least and a pretty good ultralightweight methodology at the very most. It is up to you and your team what you choose to do with the basic information you have been given!

Chapter 5: Feature-Driven Development (FDD)

Together with Crystal (which we have discussed throughout the last chapter) and the Dynamic System Development Method, the Feature-Driven Development approach (FDD) is less popular as a methodology per se.

This is largely due to the fact that FDD tends to be associated and annexed to some of the more popular methods, instead of functioning as a standalone methodology in its own right.

This is not to say that Feature-Driven Development is not important or that it cannot function on its own, on the contrary, actually. It can and you can definitely try it out.

Rather than debating whether or not FDD is popular or relevant, or whether or not you should actually implement it as a pure methodology, we want this chapter to be an informative one that will set things on a clearer path for you and help you understand not only what FDD is and where it stands in the world of agile but also how you can incorporate it in a unique and personalized approach to agile project management in general.

At its very core, Feature-Driven Development is not much different than many other agile project management methodologies. The one quality that differentiates it from the rest of the agile methods is the fact that, as the name suggests, it focuses on making progress on each feature.

Basically, the importance of the "story" delivery you know from Scrum is transferred to the importance of each feature. Sometimes, a story might coincide with a feature, but this is not a must.

Developed at the end of the 1990s for a project that aimed to deliver a short time-spanned project for the banking industry in Singapore, FDD is nowadays used quite widely. Many people automatically incorporate it in their Scrum or Kanban approaches, but it is important to note that just because this is common, it doesn't mean FDD and Scrum are one and the same (or that FDD and Kanban are, for that matter).

Feature-Driven Development is one of the more prescriptive agile methodologies out there in the sense that it works based on a clearly defined life cycle, and it assigns clear roles among the different team members.

The FDD life cycle is defined by five main stages at which the product is developed:

1. Developing the overall model. At this stage, the team gets familiarized with the high-level scope of the entire project and system, as well as the context it comes from. Further on, the team will split into smaller groups, and each area of the system will be modeled then presented for peer review. The model(s) that are deemed a better fit will be

selected to act as a domain area model. In time, all the domain area models will be merged into the larger model.

2. Building a feature list. Once the knowledge has been collected throughout the first stage of the development, all the information will be used to identify the list of features. The domain area will be split into multiple subject areas, according to the functionality of the features representative of the area. It is important to note that each feature should be identified through the prism of the value it provides the user with.
3. Planning by feature. Once you have the entire list of features at hand, you will be able to start planning the development per se. During this process, you will assign the feature sets as classes to the programmers in your team.
4. Designing by feature. Each feature of the product comes with a design package that will be handled during the fourth stage of development. This will be handled by a chief programmer in a team, who will initially select a set of features to be developed over the course of two weeks. Together with the class owners associated with these features, the chief programmer will create sequence diagrams for each of the selected features and use them to refine the model.
5. Building by feature. Once the prework has been done and once the design inspection is ready, it is time for the team

to move on to the actual programming, then test and inspect the code to ensure the feature is complete. Once that happens, it can be incorporated into the system.

As for the roles encountered in FDD, they are quite clear, and they may include the following:

- Domain Manager
- Language Guru
- Build Engineer
- Release Manager
- System Administrator
- Tester
- Technical Writer

In some respects, FDD and XP are quite similar to each other. It is very important to note, however, that the major difference between the two comes with the introduction of a "class ownership" concept. As mentioned in the third chapter of this book, collective ownership is one of the particularities in Extreme Programming. In Feature-Driven Development, however, this ownership is transferred to the class owner who becomes responsible for its functionality.

The most notable tip of information you should keep in mind when it comes to how FDD functions is that instead of focusing on more or less random chunks of the project, it approaches development on a feature-by-feature basis.

Same as in the case of the other agile project management methodologies, Feature-Driven Development is associated with a series of best practices too. We will tackle them in the remainder of this chapter, one by one.

Domain Object Modeling

Domain Object Modeling is one of the most important best practices associated with FDD. This practice basically consists of both exploring and explaining the domain of the problem at hand. Once the domain object model is generated, the team will have an overall framework to use when adding features, one by one, as they are developed.

Developing by Feature

As mentioned before, this is one of the main tenets of Feature-Driven Development (as the name of this methodology suggests a well). There is one important rule associated with the concept of developing by feature: if a function is too complex to be developed and implemented in two weeks, it will be split into smaller functions until each of the resulting subproblems is small enough to be considered a feature in the full sense of the word.

This allows for better control of the changes that might appear along the road, be they related to changes in product requirements or changes related to bugs and/or lack of functionality.

Component/Class Ownership

This best practice was also touched upon in the introductory section of this chapter, but it is quite important to keep it in mind, so we will mention it here as well. What class ownership means is that different pieces or groups of pieces of code (called "classes") are assigned to specific owners. Each of these owners is responsible for the code on multiple levels:

- consistency
- performance
- conceptual integrity

Feature Teams

Same as most agile methodologies, Feature-Driven Development prefers working with small teams. As such, feature teams will be assigned for the development of each feature.

A feature team is a small team formed dynamically to develop a small activity. Together, they work for each design decision, and they evaluate their options before they choose a particular one.

As you can see, there's quite a lot of trust placed in each and every member of these teams, as well as in how they can work togethe and cooperate for the success of the entire project.

Inspections

According to FDD's best practices, you should run regular inspections of the design and code, so that you can detect any defects in due time (and so that you can apply the changes necessary to improving these defects).

Configuration Management

Configuration management is quite important, especially when new team members join in or when you are adopting FDD for the first time. However, it is a best practice that should be maintained regardless of where in your journey to Feature-Driven Development you may be.

Basically, configuration management will allow you to identify the source code for everything (all the features) that have been developed to date. This will also allow you to keep track of the class changes while feature teams work on their enhancement.

Regular Builds

The concept of "regular builds" is similar to that of "continuous delivery" in the general Agile Principles. What this concept refers to is ensuring that your system can demonstrate, at any time, that integration errors have been fixed early on in the process. This allows the customer to maintain trust in you and your team, and it allows you to actually keep track of all the issues, whether or not they are repeat offenders, and whether or not they have been successfully fixed.

Visibility of Progress and Results

This concept is quite similar to the Scrum and/or Kanban board in the sense that the progress and the results of your team's efforts should be visible at all times. FDD may not employ an actual board for this, but the FDD project manager needs to ensure that all reports are frequent, accurate, and appropriate both from the internal point of view (team, internal stakeholders, higher management, etc.) and from the external point of view (customer).

Feature-Driven Development is, as mentioned in the beginning of this chapter, a less common methodology on its own. However, it is quite essential to remember that it can be easily integrated with other agile methodologies.

In fact, the first two stages of the FDD development cycle are almost entirely congruent to the initial envisioning model of Agile Model Driven Development, showing that FDD belongs in the agile family just as much as Scrum, Kanban, or Extreme Programming.

Indeed, there is less focus on people in FDD (as compared to Scrum, Kanban, XP, or Crystal, for example), as the main center of this approach lies on the actual feature and how the *people* around it help with its harmonious development.

However, this doesn't make Feature-Driven Development any less agile in nature. At the end of the day, FDD abides to the general Agile Principles just as much as any other methodology in the book.

It is easy to dismiss the lesser popular agile methodologies out there, including FDD, especially since there seems to be less information and a smaller number of tools that are designed specifically for this approach.

However, what you do have to keep in mind is that methods like these are more of a supporting system to the more popular methodologies that focus on *mindset* more than *specific approach.*

FDD can work in combination with most of the project management methodologies we have described so far, with the exception of Extreme Programming (where the concept of class ownership and that of collective ownership will clash).

Same as in the case of Crystal, we thoroughly encourage you to learn more about FDD as well. Aside from the five main stages we have described here, each of them is associated with specific substeps that will allow your entire plan to be more structured. For this reason, FDD is one of the agile methodologies that seem to be more of a better fit for very structured businesses (like large corporations, for example).

As we have emphasized throughout this book (as well as the first installment of the series), we thoroughly believe in the fact that every business should find their own agile path. For some, the "traditional" Scrum or Kanban may be enough. For others, however, a more complex combination might be needed. Experiment and see what work for you!

Chapter 6: Dynamic System Development Method (DSDM)

At this point in the book, you have learned about the two most popular agile project management methods (Scrum and Kanban), the second most popular one (XP), as well as a couple of methods that seem to have been forgotten but which can prove their value, especially in a hybrid context (Crystal and FDD).

For the last chapter of this book, we have decided to dive a little deeper into a method that seems to be very complex (and thus, quite scary especially for beginners in the art of agile project management): Dynamic System Development Method (DSDM).

If in the case of FDD the name is quite suggestive of the nature of what the methodology employs, DSDM may appear to be a total enigma when you first look at it.

We understand why it might sound downright confusing and why you might not be that open to learning about it if you are at a beginner or even intermediate level of agile project management. However, we must mention that DSDM is far easier than it sounds.

Like all the methodologies we have approached in this book, DSDM too abides by the general Agile Principles. As a result, it too can be defined as an agile, iterative, and incremental project management methodology. Like many of the other methods we have described here as well, the Dynamic System Delivery Method started off as a framework pertaining to the world of software programming. In time, however, some of its concepts and principles have transferred to other industries as well.

The DSDM approach has been largely developed upon the foundation laid by RAD (Rapid Application Development), a method that lies at the confluence between agile, adaptive, spiral, and unified project management. In essence, RAD is another way of approaching agile project management, but one that comes with its own set of principles and specificities that make it look like it pertains to everything nontraditional in the field of project management in general.

The delimitation between RAD and agile are very hard to make, and, as such, most of the theoreticians are happy to include RAD (and its offspring, including DSDM) under the wide umbrella term known as agile project management. It's not wrong to do this, as the principles are congruent, and there is a common antitraditionalist factor to the entire point of view in both the case of agile project management and Rapid Application Development.

In many ways, RAD represents just a slice of agile when you compare the core of the methodology. It focuses a lot on the speed of production and on delivering working software, and it doesn't focus on procedures and documentation.

What RAD lacks as compared to the other agile methods is the focus on the human element of a development project, as well as the mindset behind it. In this respect, RAD and FDD are quite similar, but it is important to note that DSDM has come as a solution to this because it includes the human element in the entire process and places it at the center of its main principles.

Moving to the actual topic of this chapter, Dynamic System Development is a methodology that sets out the time, the costs, and the quality of the final delivery from the onset of the project. In order to make sure it sticks to the plan, this method splits tasks into four main categories according to their priorities:

- musts
- shoulds
- coulds
- won’ts

One feature of DSDM you might find interesting is that its handbook is available for free, online. Furthermore, multiple resources and templates are also available for download, for those of you whose curiosity might be stirred after reading about the basics of DSDM.

Before we dive into the specifics of Dynamic System Development, we would like to take a moment to analyze its eight main principles: the guiding light of the entire methodology and the core ideas according to which everything DSDM happens.

In short, the DSDM principles you should be familiar with include the following:

- Business needs are very important. It is crucial to understand the specific business need behind the project to be able to deliver what the customer needs.
- The delivery must be made on time, as this helps keep the project on its right course and allows you to keep the customer happy.
- Collaboration is a core value of the DSDM approach, same as in the case of all the other agile project management methodologies out there. You should be able to properly collaborate as a team, as well as with external and internal stakeholders (including the customer).
- The quality of the product should never be compromised. At the end of the project, this is what brings business value both on your end and on the end of your customer, so it is essential to make sure the product abides to the highest standards of quality possible.
- You should build incrementally, but it should all start with a proper, solid foundation. Without that, the "house" will crumble, so take your time in ensuring that the foundation is solid enough for you to start building on.
- You should deliver iteratively, same as with all the other agile project management methods. This allows constant feedback to be properly incorporated into the development

process, so that the final product is suited to the client's needs in all respects.

- Communicate at all times. You should be able to communicate clearly and continuously, both internally, as a team, and externally, with your customers and/or other external stakeholders.
- Demonstrate that you have control over the entire process. As a DSDM adopter and as an agile practitioner, you should be on top of the changes that come along the way, and you shouldn't allow them to spiral out of control.

Furthermore, aside from the core principles behind Dynamic System Development, you should also keep in mind that this method is characterized by a series of specific techniques one should use. They include the following:

- Timeboxing. This concept is quite interesting because it will provide you with a mindset that will help you prioritize tasks. In DSDM, the time needed for delivery and the budget needed for delivery are fixed variables. As such, the only flexible variable you are stuck with are requirements.

  This means you should prioritize the requirements according to the importance they have in the overall functionality of the product. If you run out of time or money, you should leave the least important features out of the iteration and deliver the working product as such.

As long as the essential 20% of the requirements are satisfied, you can consider your iteration delivery a successful one.

Do keep in mind that this does not mean you are allowed to deliver an unfinished product. However, the requirements left out of an iteration should be seen as a way to improve the product throughout the future development process.

- MoSCoW is an acronym for what was already mentioned in the beginning of this chapter: work items and/or requirements can be prioritized according to whether or not they *must* be done, *should* be done, *could* be done, or *will not* have to be done.
- Prototypes are an important concept in DSDM, as this method advocates for the creation of system prototypes early in the development process. This will allow you and your team to discover any kind of shortcomings your system may have and do it early in the development process, so that amends can be applied as soon as possible.
- Testing is another crucial concept in DSDM (one that is common to most of the agile project management methodologies, actually). Continuous testing and feedback allow for the creation of a better product in the end.
- Regular workshops that bring together the different stakeholders of a project are also encouraged. These

workshops allow the stakeholders to discuss the requirements, the functionalities, and the specifics of the project and ensure that everything is crystal clear for everyone involved.

- Modeling is a concept that is somewhat common to FDD but with a lack of emphasis on feature modeling and more of a focus on visualizing business domains so that you can improve your understanding of the product and its requirements.
- Configuration management is a practice that helps managers handle multiple deliveries at the same time, at the end of each timebox.

As you have probably noticed thus far, DSDM is a pretty prescriptive approach in the world of agile project management (at least as compared to other methods in the same spectrum). Thus, it makes sense that it comes with a set of predefined team roles as well:

- The Executive Sponsor (also known as the "Project Champion") is on the side of the user organization and has the power to divert funds towards the development of the project.
- The Visionary is the person who initializes the project and draws the basic requirements earlier on in the process.

- The Ambassador User is the person who connects the user community and the project and makes sure that essential information reaches the development team.
- The Advisor User is a user that is actively involved in providing feedback and ideas with relation to the product being developed.
- The Project Manager is either a user or a member of the production/IT Team who is assigned to the management of the project.
- The Technical Coordinator is a person responsible to create and design the system architecture and ensure the technical quality of the product.
- The Team Leader is a person whose main purpose is that of leading their team and making sure that they work effectively and efficiently.
- The Solution Developer is a person charged with the interpretation of the system requirements and their modeling.
- The Solution Tester is the person responsible for checking the correctness of the code, as well as raising the issue when defects are spotted.
- The Scribe is responsible for the collection and recording of the requirements, agreements, and decisions of every workshop. In general, another member of the team will be temporarily assigned to this task.

- The Facilitator is the MC of the workshop—the person who makes it happen and ensures that discussions stay on track.

In addition to these roles, some DSDM projects will also include specialist roles, such as that of the Business Architect or System Integrator. These roles are not prescribed but recommended under certain circumstances, depending on the complexity of the product and on the complexity of the project in general.

It is also quite important to mention that the DSDM method also ascribes certain factors to the potentiality of success. In short, there are four main points that make a project successful:

- Senior management must accept and support the implementation of DSDM as a project management method. This helps ensure everyone is motivated throughout the entire duration of the project.
- End-user involvement is crucial as well, and for this to happen, you must make sure that higher management is willing to get them involved as well.
- The team must include skillful members. But maybe even more importantly, the team must be constantly empowered, not just in the sense of putting them in the right mindset but also in the sense that obstacles should be removed from their way. For instance, the team should not ask for approval for every small change. Instead, a

delegated person in the team should be empowered with a certain level of decision making. This way, time and effort will not be wasted by running paperwork through the management chain of the organization.

- Customers and vendors should maintain a supportive relationship based on communication and collaboration.

In very brief terms, this is what DSDM is all about. Combined with the specific practices of this method's lifecycle (which we will discuss further on in each of the following sections), these principles and rules make for what is generally known as the Dynamic System Development Method.

As you have probably noticed, most of these concepts are congruent with agile project management. So, if you are familiar with its general ideas and Principles, you will find that internalizing the innerworks of DSDM is quite easy.

Let's take a closer look at the Dynamic System Development Method lifecycle and everything it entails.

Feasibility and business study

The first stage of the DSDM lifecycle consists of two main phases: the feasibility study and the business study. They have to happen in this specific order because the feasibility study will provide you information for the business study as well.

During the feasibility study stage, you should study how feasible (or how possible) your application idea is. If it is deemed feasible according to specific criteria, you will have to look into the available team and the available budget. At the end of the stage, you will have to prepare a report on how the product meets the feasibility criteria (time, budget, resources, etc.), as well as a prototype (model) of the project.

During the business study stage, the business experts and the technical experts are brought together to discuss the main issues that might arise throughout the development of the product. The problems will be listed and documented for further reference. Together, the business and the technical experts will determine if they have the business and the technical capabilities to handle the project successfully. If the result of the discussion is a "Yes," the meeting (or series of meetings) will be concluded by a list of requirement priorities, as well as diagrams of the application and product infrastructure.

Functional model/prototype iteration

Once you have all the requirements set in place, the data collected during the predevelopment phase (the first stage of the DSDM lifecycle) is pulled together in a functional prototype of the product. The model will include all the requirements, and it will organize them incrementally.

The prototypes will be further studied and split into smaller substages, including:

- The identification of the functional prototype—the key functionalities you want to include in the prototype.
- The creation and acceptance of the plan and of the schedule.
- Creating the functional prototype: bringing in the programmers and asking them to create a prototype of the product based on what has already been identified and planned in the previous stages. It is important to include a testing phase here as well. Just because this is a prototype, it doesn't mean that it shouldn't be perfectly functional.
- Reviewing the prototype. Once the prototype of the product is ready, it is time to place it in the hands of the end users and ask them to test it. Their feedback and comments will be taken into consideration for future iterations, as the product prototype will continue to be grown and improved throughout the entire duration of the project.

Design and build iteration

The third phase of the DSDM lifecycle is all about building on the prototype and ensuring that it is a continuously improved product. From the prototype, you and your team will move on to developing specific functionalities (also known as individual units) and integrating them into the system.

It is worth mentioning that in Dynamic System Development, there is no clear distinction made between design and build; both of them are handled during this stage of the DSDM lifecycle.

Furthermore, it is important to mention that this stage is also split into four smaller substages, as follows:

- The identification of the design prototype, which includes the requirements that have been decided upon in the prototype/model and then prioritizing them.
- The acceptance of the plan and scheduling. Once the requirements have been planned and scheduled, they need to be agreed upon with the team.
- The creation of the design prototype. Same as in the case of the functional prototype, this stage will deal with the actual development and testing of the design prototype.
- Reviewing the design prototype. At this stage, you will allow the design prototype to be tested and ensure that it is correct and congruent to what was initially planned. Any changes will be implemented in future project iterations until the product is ready for a full release.

Implementation

If the previous stages of the DSDM lifecycle were more about planning and developing small pieces of the product for the purpose of observation, this stage is all about actual implementation.

This is all about watching the live action unfold: putting the product in the hands of the end users and allowing them to fully test it and see if all the business requirements are met.

The Implementation phase is (perhaps unsurprisingly) split into four substages as well:

- Getting the user's approval and offering them guidelines on what should be tested more specifically
- Training users
- Implementing the feedback, you receive from the users
- Reviewing the business needs and whether or not the product meets them. If any new valuable features are identified, they will be further implemented in the product.

Like all agile project management methodologies, the Dynamic System Development Method is based on multiple iterations, so the steps will be repeated for as long as necessary to ensure that the product satisfies all needs and that it is fully ready to hit the market.

As we were saying at the beginning of this chapter, DSDM might sound overly complex at first, but once you nail down its basics, you will understand that this specific method is, just like FDD, a very suitable one for organizations that are more focused on structured planning and documentation. For you as the DSDM project manager, the emphasis will lie on delivering working products. For your organization, however, DSDM and FDD can provide you with plenty of structured information and reporting to keep higher management sound asleep at night.

As you have seen, all agile project management methodologies are more or less variations on the same topic. All of them abide by the Agile Principles. And all of them aim to deliver quality, timely, and budget-friendly products—in software development and every other industry that has embraced agile at the core of their functionality.

Conclusion

In 1971, Ray Tomlinson sent an Earth-shattering message: *something like QUERTYUIOP (*Computing History, 2019).

OK, the message itself was pretty far from life changing. It was quite nonsensical, in fact.

What that message was doing, however, was about to shake the world. Nearly five decades later, we wish that message would have been something more epic, like *a small message for man, a big message for mankind*. But it wasn't. It was just like all those random tests people run when they implement a new system. They make no sense, they come out of the blue, and some of them get to write history in some of the most ridiculous ways possible.

Tomlinson's message was exactly that: a ridiculous experiment that entered history. The message behind the letters written on the screen doesn't even matter today. What matters is that this was the first email ever sent, and it set in motion a series of changes that affect us to the date.

Can you imagine your coffee break without a scroll on Facebook? Can you imagine arguing with your friends on what year the first Terminator came out in and not being able to get on Google and find they were right? Can you imagine not posting photos of your wedding day for all those envious high school classmates to see?

Probably not.

The internet has grown to be such a massive part of our lives that it is impossible to imagine life without it. At work and outside of work, we are constantly relying on the internet to provide us with the connection we need. We rely on it to connect us to information, entertainment, and downright bitterness over the decades that have passed since our high school graduation.

What Tomlinson did mark the beginning of a whole new era in the evolution of the internet. Sent via ARPANET (a primordial ancestor of modern-day internet), his totally nonsensical message was, in fact, the beginning of online communication as we now know it.

It took another couple of decades until the internet got in every home, but beginnings are always timid like that. Once set in motion, the machinery powering the bytes behind the screen blew up to an entire phenomenon, one we now know as the (in)famous dotcom bubble.

On this background, software development companies were facing one of the most gruesome and most irritating issues in the world of tech: lag. They were lagging so much behind schedule that some of them delayed product releases by *years*.

The real issue was not that there wasn't enough talent or that these companies were lazy in any way. The real issue was connected to the fact that all these projects were very volatile in nature. Requirements were changing all the time, new technologies were emerging all the time, and the very nature of these programmers' jobs was changing from one day to another.

And yet, they were all stuck with strict management methods that had absolutely no idea how to deal with the massive amount of uncertainty dominating the software development world.

Everyone was in dire need of change. They needed a framework that would be able to actually help them respond to all these shifting situations.

One by one, lightweight agile methods started to pop here and there. Crystal, Scrum, and XP were among the first ones. And before anyone realized what was going on, the Japanese came from the far East to put on the table a solution they had been using for several decades by then: Kanban.

From a world of strict rigor and nearly obsessive planning, the world of software programming was blown up by the chaos of all these innovative, albeit unregulated new methods.

And then, in 2001, a bunch of smart guys met up for a ski trip, sat down, and laid the foundation of the Agile Manifesto: a framework that brought together all the lightweight methodologies and unified them under a common set of Principles.

Magic history is rarely planned. The first email sure wasn't, and it is more than likely that the creation of the Agile Manifesto wasn't a thoroughly planned moment either. There were no confetti popping in the background of these moments. No fireworks. No special concerts. Nothing to say that those were special days in the evolution of mankind.

And yet, these moments happen. They happened behind locked doors and as a result of hard work and determination and frustration. They happened because it was the only natural thing to do. They happened. And they changed the world.

By the end of this book, we truly hope you have come to grasp the full agile project management grandeur in all its wonderful formats and colors. We hope you have come to understand that good planning is not about knowing what will happen, but about knowing how to respond to what you don't know yet.

The best moments in mankind's history are moments of shifting change, moments that happen silently, without much buzz around them. What would we be as a race if we didn't know how to adapt to these changes?

You see, agile is the most natural thing in the world. Mankind has been inadvertently doing it ever since we started using tools and making fire. And then, we somehow stopped doing it naturally and started closing ourselves and our resilience in the face of change behind strict rigors and obsessive-compulsive planning.

Agile project management is one of the most natural paths you can take, especially in the ever-changing environment of today's world. It's not even a matter of adapting to the slow changes that happen behind locked doors anymore: it is a matter of adapting to massive changes in everything around us.

From climate to politics to society, our world is drifting apart from its traditional views and opening the gates to a whole new future—one only science fiction writers could have ever imagined.

The wonderful thing about everything we have presented in this book (as well as the first installment of our Agile Project Management series) is that *agile* has been the motor of change for quite some time, and it will continue to drive change as more and more companies from wildly varied industries start embracing it.

The more flexible you become, the easier it is for you to truly embrace what comes ahead, and agile project managers know this better than anyone. Being inflexible, on the other hand, will only lead to broken bones and painful failure.

We hope this book was able to provide you with a full, comprehensive overview of some of the most popular agile project management methodologies in the world. We do not claim we have said it *all*. It would be completely absurd for us to say that. Instead, we claim that our book is one of the best pieces you will ever read on the topic of how the main agile methodologies connect into their agile mothership, and how they connect to each other.

As we said in the introduction, we cannot tell you the *right* answer. Agile may or may not be for you because yes, it is a perfectly valid point that agile is not for every single company out there.

What we do hope is, however, that you will give agile a chance. Even in its strictest forms and even in hybrid formats, agile can still be the internal change you need to be able to withstand the external shifts in the world.

Our journey through Scrum has shown you that there is a very good reason this method is very popular but that you should look beyond the apparent playfulness of the practices employed in this approach as well.

Our journey through Kanban has shown you that something can be old and new at the same time, and that agile does not always originate in the realm of software programming.

Our journey through Extreme Programming has shown you that sometimes, popular methods fall back into near oblivion, but they continue to survive through what they do best: the specific practices that have been borrowed by other project management frameworks.

Our journey through the Crystal Method has shown you that some of the largest companies in the world can become agile too, despite all the apparent challenges they had to face in the process.

Our journey through Feature-Driven Development has shown you that it doesn't matter how complex an agile method may look from afar: when you strip down, it's all about the flexibility.

Finally, our last journey through the Dynamic System Development Method has shown you that yes, agile can be documented and reported as well. So yes, it can definitely work in organizations that treasure these features too.

Overall, we hope that this entire journey through the agile project management methods of the current landscape has shown you that there is no right or wrong solution, and that most of the times, the right path lies in knowing which practices are specifically good for you, your team, and your business.

Agile project management can be genuinely fascinating when you know how to look at it. It might sound like the dullest topic in the world when you work outside of it, and it may seem like it's an endless discussion on which spreadsheet formula works best for reminding people that they have a deadline to attend.

But when you look past all these misconceptions, you will discover that agile project management is all about the marvelous change we have all witnessed in the past few decades.

We truly believe we are not exaggerating when we say that agile has been the engine of change. Behind the great ideas of Steve Jobs and Mark Zuckerberg, behind the recent launch of a car in space by Elon Musk, and behind the most proficient businesses in today's world, agile works its way in and out of trouble. *In*, because it doesn't fear change and issues in the process. And *out* because it is the very core of the problem-solving mindset, we all need to adopt.

At this point, it doesn't even matter if you are a project manager, an entrepreneur, or simply someone who wants to achieve the best version of themselves.

All that matters are that you can take the agile methods we have described here, nitpick them, test them out, and see what works best for you. We guarantee that you *will* find the right formula to success, regardless of whether that is all about delivering a working software program or growing a pair of triceps by next summer.

Agile can change you inside and out because it shifts your entire perspective on life and how you should react to the changes around you. And for this reason, we cannot be anything but proud to say that yes, *agile rules the world*.

We wish you luck in your future endeavors, and we truly hope this book has been of help for you, for your team, and for your business. It won't be an easy journey ahead, no matter what you decide to do next, but we truly hope that this book has taught you that resilience, hard work, and pure determination can make even the most agile-adverse situations turn around and smile back.

At the end of the day, agile is all about continuously iterating your product's releases until you get it right and until you find the (perhaps not so secret) formula to success. And your very journey into agile should be a metaunderstanding of everything you have learned about this project management mindset and everything it brings along!

Stay strong, stay smart, stay curious, stay agile, and the Universe will smile down upon you!

References

Cherry, K. (2019). The cognitive and productive costs of multitasking. Retrieved 9 October 2019, from https://www.verywellmind.com/multitasking-2795003

Emmons, M. (2019). Key statistics about millennials in the workplace. Retrieved 9 October 2019, from https://dynamicsignal.com/2018/10/09/key-statistics-millennials-in-the-workplace/

First network email sent by Ray Tomlinson. (2019). Computing History. Retrieved 9 October 2019, from http://www.computinghistory.org.uk/det/6116/First-e-mail-sent-by-Ray-Tomlinson/

Ropa, S. (2015). Is extreme programming no longer relevant?. Retrieved 9 October 2019, from https://resources.collab.net/blogs/is-extreme-programming-no-longer-relevant

CPSIA information can be obtained
at www.ICGtesting.com
Printed in the USA
LVHW010152220221
679513LV00003B/248